Natural Science in Archaeology

Series Editors: B. Herrmann · G. A. Wagner

Springer
Berlin
Heidelberg
New York
Hong Kong
London
Milan
Paris
Tokyo

Ervan G. Garrison

Techniques in Archaeological Geology

With 102 Figures and 20 Tables

 Springer

Author

Professor Dr. Ervan G. Garrison
University of Georgia
Department of Geology
Athens, Georgia 30602-2501
USA
E-mail: *egarriso@uga.edu*

Series Editors

Professor Dr. Bernd Herrmann
University of Göttingen
Institute of Anthropology
Bürgerstraße 50
37073 Göttingen, Germany
E-mail: *bherrma@gwdg.de*

Professor Dr. Günther A. Wagner
Max Planck Institute of Nuclear Physics
Institute of Archaeometry
Saupfercheckweg 1
69117 Heidelberg, Germany
E-mail: *gwagner@goanna.mpi-hd.mpg.de*

ISBN 3-540-43822-X Springer-Verlag Berlin Heidelberg New York

Cataloging-in-Publication Data applied for

A catalog record for this book is available from Library of Congress.
Bibliographic information published by Die Deutsche Bibliothek.
Die Deutsche Bibliothek lists this publication in the Deutsche Nationalbibliografie:
detailed bibliographic data is available in the Internet at http://dnb.ddb.de

Springer-Verlag Berlin Heidelberg New York
a member of BertelsmannSpringer Science+Business Media GmbH

http://www.springer.de

© Springer-Verlag Berlin Heidelberg 2003
Printed in Germany

Production: PRO EDIT GmbH, Heidelberg, Germany
Cover Design: design & production, Heidelberg, Germany
Typesetting: SNP Best-set Typesetter Ltd., Hong Kong

Printed on acid free paper   32/3141/Di   5 4 3 2 1 0

Acknowledgements

The author would like to thank the many people who contributed to the final form of this work. All faults and factual errors are those of the author alone. The kind and critical attention so many colleagues paid to the drafts of this book is greatly appreciated.

I am grateful to the following individuals: the Series Editor for Springer-Verlag, Günther Wagner, Gilles Allard, John Dowd, Norman Herz, Steve Holland, David Leigh, Phil Reppert, Sam Swanson and David Wenner. Other colleagues generously shared insights and materials which enhanced many illustrations. These include: George Brook, Mike Douglas, Ken Kvanne, Sherri Littman, Mike Roden, Kent Schneider, Nina Šerman, Paul Schroeder, Paul Sinclair, Don Thieme and Wendy Weaver.

Brenna J.M. Williams typed much of the earliest draft of the manuscript. Beatriz Stephens, Mary Crowe and Elizabeth Meadows of the University of Georgia's Geology Department typed a significant portion of the manuscript draft.

The author is extremely grateful to these women for their expertise and patience. The drafting of many key illustrations is the work of Wendy Giminski of Campus Graphics and Photography, University of Georgia. Production editing was done by Ms. Luisa Tonarelli of Springer-Verlag.

Finally, the author thanks Ginger Garrison for the nights and weekends of support and encouragement. Merci à tout le monde!

Contents

1 Introduction

This book is a survey of techniques used in archaeological geology. It is less a discussion of theory or methodology with regards to the various geological techniques that are presented. It is not an exhaustive presentation of the diversity of earth science methods that can be utilized in the service of archaeology. In many ways it is a personal view of how earth science can be used in archaeology. The author is perhaps more archaeologist than earth scientist, so that the archaeological "ends" have always determined the geological "means" in studies over the past decades. Certainly, there are other ways to approach the multiplicity of archaeological questions that confront the student of the past and this book does not presume to suggest that the techniques presented herein are the only ones appropriate for addressing the many facets of archaeological inquiry.

The acceptance of earth science techniques into archaeological study has been gradual, but has grown to a point that we now have university-level programs in archaeological science, geoarchaeology and/or archaeological geology. As Norman Herz, in *Geological Methods for Archaeology* (Herz and Garrison 1998), stressed, it is less what one calls the use of earth science techniques in archaeology than what one realizes from their application. The practice of modern archaeology demands an understanding of earth science. One does not have to look any further than that of the importance of first, Sir John Evans and Joseph Prestwich, then Sir Charles Lyell's 1859 visits to the gravel beds of the Somme River. In the Somme gravels, Boucher de Perthes painstakingly demonstrated the contemporaneity of ancient stone artifacts and extinct paleontology. Their visits brought newly developed insights in the science of geology to the interpretation of the Somme deposits and greater scientific validity was a result. This interplay of archaeology and geology can result in similar advances in our understanding of our human past.

An appreciation for the utility of geology for archaeology developed among many archaeologists in Europe and the Americas in the twentieth century. In the USA, for example, the use of geological methods at the Midland site by Wendorf and his coworkers in the 1950s set an important and enduring example of how geology and archaeology buttressed each other to mutual advantage (Wendorf et al. 1955). Often, however, where geology was included in a site report, there was a lack of an integrative framework that would have incorporated the geology into the overall explanatory perspective of the research Awkward sentence. In many field reports, the geological section is more often referred to by the author(s) in a meaningful way. In the mid- to late twentieth century, many innovations were made in the *routine* of excavation. Instrumentation and technique, in particular, have been the most affected by revolutions in digital technology, remote sensing and analytical techniques. Most digs are carried out today with the presence of computers and their peripherals together with devices that use the "p.c." for data storage and manipulation. The explication of geoarchaeological techniques is important, since they are more and more commonplace on modern excavations. This we shall attempt to do in some detail.

The ability to synthesize several lines of scientific data is the standard by which we train students, particularly those in graduate school. It is not unrealistic or revolutionary to call for a similar methodological approach in the practice of archaeological geology. In this volume, a format of inquiry and reportage that has merit is laid out. This textbook simply reinforces similar calls by others within the geology of archaeology. These include Karl Butzer, George (Rip) Rapp, Vance Haynes, Jack Donoghue, Brooks Ellwood, Reid Ferring, William Farrand, Joel Schuldenrein and Julie Stein to name but a few of these scholars. Rapp, in his books written with John Gifford (1985), and Christopher Hill (1998), together with works by Davidson and Shackley (1976), Gladfelter (1977), Waters (1992), Stein and Linse (1993), Bettis (1995) and Pollard (1999) have provided clear models for geoarchaeological inquiry and interpretation.

Archeological geology is perhaps more "method-oriented" than methodology in that it is clearly a technique-driven enterprise in the service of both archaeology and quaternary studies. As a consequence of this, archaeological geology has no great body of theory to govern it's conduct. Nor does it have a pantheon of past practitioners such as older disciplines like physics and chemistry can present (Garrison 2001). As Gladfelter, in his acceptance of the Geological Society of America's "Ripp Rapp Archaeological Geology Award in 2001 implies, it is an honorable enterprise nonetheless (GSA Today 2002). In past years this observation has been made of archaeology itself (Daniel 1967). The application of perspectives and methods born in disciplines like geomorphology now are recognized by archaeologists as crucial to the understanding of site location, function, duration, and like questions. Over the latter part of the twentieth century the incorporation of the methods of other areas of earth science – geochemistry, geochronology and geophysics – have expanded the reach and utility of archaeological geology.

As Kraft notes in a review of archaeological geology, the workers in this area are becoming more aware of the nuances of the historical and classical record (Kraft 1994). Legend, oral tradition, mythology and the body of cultural knowledge for a prehistoric culture are being used in the search for information about ancient events – paleoenvironmental and paleogeographical. This is the essence of a "multidiscipline" like archaeological geology – use of geological and analytical techniques in concert with archaeological expertise to gain knowledge of the past. This book echoes many of these tenets set forth by these other workers. In so doing, this work will serve to amplify that call for technical and methodological rigor in the use of earth science in the practice of archaeology as a whole.

1.1
Organization of This Volume

This book is comprised of eight chapters. The contents of these chapters cover major elements within archaeological geology. One significant aspect of earth science that is omitted is that of geochronology. This was consciously done. Given the present-day literature on the use of dating techniques in archaeology, another survey of this area hardly seemed warranted. The reader is directed to the premier volume of this series by Springer-Verlag, *Age Determination of Young Rocks and Artifacts* (Wagner 1998). The focus of the book will be on a systematic approach to the archaeological site, its features and artifacts from an earth science perspective. While geochronological techniques are central of the understanding of stratigraphic contexts of sites, their in-depth discussion will remain outside the scope of this text. What remains are discussions of those areas of earth science that bear directly on the nature of archaeological deposits as interpreted from the geological perspective.

Chapter 2 will address methods used in the mapping of the geomorphological and geological context of the archaeological site. Immediately obvious is the question of scale to be considered by the researcher. This devolves, in turn, on the archaeological objectives of the specific research. Regional scale mapping would be called for by a research design that seeks to determine the spatial context (location) of a cohort of sites, their cultural and temporal placement important, but is secondary at this stage of the inquiry. Below regional scale would be the local environment of the individual site. This is, perhaps, the most familiar type of geological mapping used in archaeology. From a geomorphological standpoint, the identification of landforms is important. From the geological standpoint, the identification of structure and the characterization of the area's lithology from outcrops and exposures must be done as well. To do this the archaeological geologist relies on traditional as well as contemporary means.

In the area of traditional geological mapping, the researcher relies on their training in landform and soil recognition; basin hydrology and structural geology. The basic tools are maps – geographic, geological and hydrological – together with the compass, the inclinometer, the shovel, the rock hammer and the hand lens. Today there are newer tools such as satellite images, satellite location or global positioning system (GPS), laser transits and levels, mechanical excavation, coupled with laptop computers and geographic information systems (GIS). Whatever the technique or tool, the aim is the same: characterize the geomorphic and geologic context of the archaeological site or sites. Once this is done the archaeologist has a better understanding of the "why" of the site's location in a particular landscape. The past landscape or landscapes generally are masked by modern earth system processes such that the determination of the prehistoric landscape becomes an archaeological inquiry in and of itself.

Chapter 3 deals with the geophysical techniques most prevalently used in "shallow" geophysics and the most applicable to the study of archaeological features. Shallow geophysics is a modern outgrowth of the need for techniques that can characterize near-surface geology together with environmental pollution. This is a geophysics that is basically nonseismic in nature or one that does not use sound as the principal means to measure subsurface rock and sediment thickness. When applied to archaeological problems this form of geophysics has been called "archaeological geophysics" or more simply "archaeogeophysics". Weymouth (1996) has called this use of geophysics "digs without digging". Beginning in the 1960s with the application of electrical methods such as resistivity profiling, the field has expanded with the addition of electromagnetic techniques such as conductivity meters and metal detectors; the use of magnetics in the form of magnetometers and gradiometers and in the 1980s and 1990s, the use of ground penetrating radar (GPR). Coupled with computer software that is basically derivative from seismic-based geophysics, the expansion of geophysical prospection into archaeology has produced a body of research that is recognized within both archaeology as well as geology.

Chapter 4 examines specific ways to carry out more detailed investigation into the sedimentological context of the archaeological site and its environment beyond that of the present-day or near-surface components. The prehistoric landscape can, and oft-times is, more deeply found than what is capable of recovery with the shovel or even mechanical excavation. The recovery of sediment columns from hand-operated and mechanical coring devices is a favored means of geoarchaeological study and has a successful history in application. Cores produce not only sediments, but macrobotanical, macrofaunal and artifactual remains as well. Cores taken in a systematic campaign across a site, and its surrounding area, can yield significant amounts of environment, depositional history and spatial extent. Mechanical excavation, in exposing soil and sediment profiles, are superior to coring for obtaining the "big picture" as regards the geomorphic mapping of the lateral continuity of sedimentary deposits. Both hand and mechanical methods have their place in archaeological geology. It is difficult to visualize the deployment of a large mechanical excavator or even a drill rig on a prehistoric burial mound given the rarity and uniqueness of these features. Gone are the days of total excavation of these types of structures. Careful and judicious sampling of them requires methods that give the researcher greater control (and analytical returns) over the technique to be used.

Chapter 5 addresses the analytical procedures for examining sediment and soil samples acquired by excavation or coring methods. The five "p's" – particle size, point counting, phosphorus content, pollen and phytoliths – are discussed. Mineralogical identification of sediments will be discussed in the context of petrographic methods in the following chapter. Many of the methods presented are hardly "cutting edge", but they have had significant development and application across a broad range of archaeological materials achieving reliable results. For example, many researchers doing particle size analysis will find the promotion of sieves and hydrometers over more modern laser size determination methods a bit dated, but these textural analyses have great reliability and more accessibility to the average worker. Certainly, this book does not argue for any status quo, but rather an entry-level approach to standard techniques as commonly used in the field today. Things change and archaeological geology should welcome new and better (and faster) ways of doing analyses.

Laser determination of grain size is one of these new and promising methods that should draw our notice. The same can be said of a whole host of other instrumental techniques where new advances are being made each year. Field -hardened, computer-based spectroscopy is another of these areas where the lab is coming to the site more and more often. More than one geoarchaeologist sets up their thin-section lab on site and does analytical-level petrography on-site.

Chapter 6 deals with the use of petrographic methods in the analysis of rocks, minerals, sediments and archaeological artifacts or features made up of all or some of the foregoing.

The principal discussion centers on the utility of the geological thin section and the petrographic microscope in analyses of rock and mineral items of archaeological interest. Beyond the hand lens this protocol is the geologist's most reliable and time-honored methodology. As Herz (Herz and Garrison 1998) has pointed out, it is also one of the most economical and reliable ways of determining the geological nature of artifacts. It is certainly to be considered before moving to the next level of analysis – geochemical analytical techniques. What is disquieting is how often the archaeologist does not use petrography before moving to more costly methods. In the case of ceramic studies the use of petrography is becoming a major and precise method of studying ceramics and their clay sources (Vaughn 1991; Velde and Druc 1998).

Chapter 7 discusses some of the more prevalent instrumental techniques that have come to be used in the study of archaeological problems. Notable among these techniques are neutron activation analysis (NAA) and mass spectroscopy of stable isotopes. Added to these are the X-ray methods of fluorescence and diffraction (XRF and XRD). The latter method has been styled as "the archaeologist's first line of defense in mineralogical identification". Together with the electron microprobe (EMP), its high energy cousin – proton-induced X-ray emission (PIXE); inductively coupled plasma spectroscopy (ICP) and atomic absorption spectroscopy (AAS), the archaeological chemistry of artifacts and, increasingly, their residual contents – food, liquids, etc. – is the subject of major studies across all aspects of archaeology – prehistoric and classical.

In chapter 8 the statistical exploration of data derived from archaeological geological studies is presented in a very introductory fashion. Because of the need for the recognition of correlations between variables used to describe and characterize artifacts and other aspects of the archaeological sites, this chapter will examine basic bivariate procedures together with regression analysis. For multiple variables and cases above two, we shall look at simple multivariate procedures such as Q and R-mode factor analysis. Again, the objective is to expose the reader to the advantage of this approach to archaeological data more than to be an in-depth discussion of statistical methods available to archaeological geology.

This textbook in not a comprehensive treatment of the field of archaeological geology as it exists today. It is a tribute to the increasing maturity of the field as a discipline that there is such breadth and diversity in its subject matter. What I shall attempt, in the next eight chapters, is to expose the student, and any interested colleague, to practical concerns and earth science methods that are useful in the consideration of archaeological problems. Archaeology identifies the results of this interaction between humans and nature in the material remains found in the earth. Linking the trajectory of past human cultures to that of the earth system requires geological science. The synthesis of this interaction provides a diversity of roles for the archaeological geologist using the techniques outlined in this textbook.

2 Survey and Mapping the Geomorphological and Geological Context

2.1
Introduction

Until very recently archaeological geology had at its core the methods of geomorphology and sedimentology. Geomorphology is a discipline more often found in the larger field of physical geography, at least in the United States and Great Britain, while sedimentology is a specialty area of geology. While some in archaeology may wonder at a partition in what is basically earth science, the location of geomorphology within geography has had some unexpected benefits, most notably the recognition and use of the burgeoning techniques of geographic information systems better known as GIS. Likewise, geographers were quick to incorporate their traditional interest in remote sensing into GIS studies as well. Digitally based, these large-scale spatial analysis systems have proven to be a boon to geoarchaeological researchers. By integrating their questions into GIS frameworks it has been possible to converge many different lines of data into synthetic map [projections] that can be manipulated and compared so as to yield new insights into the nature of archaeological sites and their locations. Had geomorphology been housed exclusively within geology departments it is unlikely that GIS methodology would have entered the arsenal of archaeological geology as quickly as it has. Sedimentology and its close relationship to paleoenvironmental study, particularly in quaternary research, has continued to be a major component of archaeological geology. Likewise, the mapping of the surface and subsurface of an archaeological locale provides crucial information as to the why of a site's specific location and function.

2.2
Geomorphological Mapping

In Native American mythology the landscape is often characterized as either indifferent or even antagonistic to humanity. Descriptions of the ancient landscape by these and other cultures, Australian, for instance, ascribe it as resulting from the action of powerful forces often in zoomorphic form. Humanity inherited this landscape from more primal times (and species – "monsters", etc.) and it was only made more habitable by time and heroic figures such as the Hero Twins of Diné cosmology of the American Southwest. To most geologists the earth's surface is a transient feature changing in a kaleidoscopic fashion through eons. This twenty-first century view of a malleable and changeable earth surface echoes that of the ancient non-Western models. At geologic time scales, the surface-forming processes of erosion and deposition are secondary to the larger, dramatic forces of plate motion, tectonism and orogeny. Such processes operate within time intervals, outside human experience, approaching that of *Wilson Cycles* wherein oceanic basins rift open and subsequently close after millions of years (Levin 1988).

The geomorphic surface is a mixture of landforms that are an expression of (1) modern geologic processes and (2) geologic processes no longer active. Landscapes reflect present and past processes of erosion and deposition. The topography of a given landscape is the result of climate-driven processes oft-times directly coupled to tectonic activity. Rinaldo et al. (1995) presented these relationships in a mathematical model of landscape evolution which underscored the importance of tectonic uplift to the formation and persistence of geomorphic signatures. Seismicity is not a new concern in geomorphology, but its character and expression – earthquakes, faulting, volcanism – is important to unraveling the nature and origin of various relict landforms.

In the northern hemisphere, climate-driven processes have given rise to glaciations. The geomorphologist must always be cognizant of the special character of glacial landform. The interactions of landform and environment are dramatically demonstrated in North America and Europe due in large part to their glacial past. Concomitant with Ice Ages were marine regressions, and consequent transgressions, which alter fluvial systems and play much the same role as uplift in nonglaciated regions. Coastal landforms – past and present – are the expression of glacial processes acting globally through sea level change.

2.3
Geomorphic Concepts

Keller and Pinter (2002) review the basic principles used in the study of geomorphic landscapes. The most basic principle, in terms of modern, process-oriented geomorphology, is that any change in a landscape implies a change in process. Other concepts include:

1. Landscape evolution can be gradual or abrupt as thresholds are exceeded.
2. The interaction of landscape changes with thresholds result from complex processes and are termed *complex response.*
3. Various tectonic fault varieties – normal, reverse and slip-strike – are associated with characteristic landforms.

In general, these geomorphic concepts apply to that part of the Earth's history called the Quaternary. The Quaternary spans, roughly, the last 2 million years and encompasses the largest part of hominid evolution.

One of the oldest models for landscape evolution is that of William Morris Davis who articulated the "Cycle of Erosion" (1899). In this model, brief periods of uplift are followed by much longer periods of inactivity and erosion. In Davis' view the landscape had a "life cycle" from (1) youth through (2) maturity to (3) old age (Davis 1899). Young landscapes are characterized by stream incision and deep valleys followed, in mature landscapes, by less dramatic stream gradients and valley down-cutting. Mature landscapes have broad floodplains and meandering streams characterized by high discharge and sediment loads. In old age, the landscape has eroded to a planar or *peneplain* surface with low relief awaiting rejuvenation or the start of a new cycle. Davis put a time scale of a million years on his cycle – long, but archaeologically more relevant than the longer Wilson Cycles.

In this, and other process models, the landscape can and will equilibrate for thousands of years. Equilibrium is clearly a relative concept with regard to landscapes. Uplift can go on for millions of years, however imperceptively. Likewise, subsidence and/or erosion will occur over the same time scales. In many instances the interplay of these forces is a "zero-sum" situation. In the contest between uplift and erosion, the latter always wins in the Cycle of Erosion. Small wonder that archaeology is an endeavor that deals primarily with that which is buried. Yet without uplift the geomorphic surface of the Earth would be that of the unchanging surface of Mars – a true equilibrium – static – situation. Some geomorphologists define a type of equilibrium termed "steady-state" where the landscape exhibits a relative constancy over a period of 100–1000 years. This conceptual view of the landscape can be called the "average" of a land surface at this particular time scale. From a process-oriented perspective one could say the limits of equilibrium or thresholds are relatively "high", such that abrupt or dramatic landscape change is precluded at this scale.

In a tectonic sense, landscapes can be "dormant" with little actual or observable seismic activity. Generally, the most active landscapes are those associated with seismically dynamic zones such as found at crustal plate boundaries. Conversely, the most inactive areas, in the tectonic sense, are furthest from plate margins such as continental interior regions. It is easy to locate tectonically active areas. They either exhibit volcanoes or earthquakes or both. Archaeological examples come easily to mind – the Apennines; Naples/Pompeii; Thera/Akrotiri; Anatolia; the Dead Sea region; the East African Rift Valley and Pacific Rim to name but the most obvious.

The "take-home message" for the archaeologist relative to modern process geomorphology is that the geomorphic surface on (and under) which archaeology is found, is not a static surface. Even in a geologic sense the geomorphic surface changes at rates of decadal to annual scale. True, the mountain belts form at Wilsonian rates as do their valleys and adjacent basins, *but* their surfaces are modified daily, by gravity-driven processes – creep, erosion, etc. Because of this dynamic and changing surface, the archaeologist should never be lulled into thinking it is immemorial. It is not.

The landforms that we propose to study and map are the result of three factors: (1) geomorphic processes; (2) stage of evolution of landform and (3) geologic structure. To map archaeologically interesting topography one must (1) identify the geologic processes by which the landform were shaped; (2) recognize the stages of development of landform and their evolution through time and (3) recognize the topographic expression of geologic structures – dip, strike, clinal variation, etc. In geoarchaeology, the consideration of topography must include another factor – human-induced geomorphic change.

2.4
Geomorphic Setting

A brief review of geomorphic settings reminds us that there are basically only: fluvial, desert/arid, coastal, glacial, volcanic and karst/caves. Most archaeology is done in relation to fluvial landforms. Human groups have always gravitated to available water and this is readily apparent in any study of agrarian cultures. Crops need reliable water supplies and these are most often found in drainage systems. Besides available water there are rich, deep alluvial soils which ancient farmers were quick to identify and exploit. Geomorphic settings are the result of geomorphic processes – fluvial, aeolian, volcanic and glacial processes which in turn are special cases of the two primary processes: erosion and deposition. In this volume we are concerned primarily with the identification of the results of geomorphic processes rather than the processes themselves. The study of the latter is that of surficial process geology. In this chapter we recapitulate the importance of these landforms to archaeology and their attributes.

2.4.1
Fluvial Landforms

Mapping fluvial systems involves both the horizontal and vertical dimensions of the riverine topography. In creating flood plains and valleys, streams alternately downcut and deposit sediments. In montane areas alluvial fans are key features that contribute to the available sediment balance as well as the distribution of that sediment. Another key landform feature in stream valleys are terraces. It is on these topographic features that most archaeological sites are located. Determining their number and order is a primary objective in any mapping of a fluvial system. The youngest terrace (T_0) is adjacent to the floodplain and the oldest is the highest terrace (T_n).

Between these terraces the stream channel moves alternately eroding (cutbanks) and depositing (point bars) alluvial sediments, generally opposite one another. Depending upon elevation and gradient (slope), which are direct expressions of uplift, stream and valley morphology vary from region to region. The hydrologic cycle (precipitation–evaporation) determines the amount of available water to the drainage system in turn influencing the incision and growth of smaller order streams within the drainage basin. The delta represents a major fluvial landform historically significant to human cultures throughout antiquity.

Fluvial systems, along with volcanoes and glaciers, have the role of renewing "old" landscapes and building new ones. The movement of eroded sediments is principally the role of the stream. In doing so, it transports the sediment through the valley, creating terraces, levees and floodplains. Depending upon the factors of precipitation, vegetation cover, lithology and gradient, the sediment load a stream carries varies with its discharge. That latter factor is dependent on climate. For archaeological geologists the correct reading of type, texture and distribution of fluvial sediments allows the retrodiction of past climate – the controlling factor in paleo-ecologies and a primary mediating element for past human cultures.

Streams incise, creating terraces and valleys, and aggrade, creating floodplains and new channel courses. When stream base levels are lower, the stream incises even in its lower reaches. When the base level rises, due to changing climate and elevated lake/sea levels, a stream aggrades. An incising stream is more of a conduit for sediment. Only when the discharge increases to flood levels does the sediment escape the channel to deposit new sediments in the floodplain. A stream carrying large volumes of sediments, such as an outwash stream, can fill its channel to the extent that an arbitrarily higher base level is established. Streams can then regularly overflow and significantly increase the volume of sediment in the valley. Soils formed or forming on the valley floors are buried, becoming paleosols, often together with archaeological facies associated with these land surfaces. Streams, in incision, and meander, likewise, can exhume past land surfaces. An excellent archaeological example is a discussion of alluvial stratigraphy and archaeology in Africa (M.A.J. Williams 1984).

Weathering of the uplands, coupled with and factly caused by precipitation, is the ultimate source of the basin sediments. The sediments are thus a direct record of their parent lithologies and the processes active on these rocks and minerals. For instance, if we observe resistant materials, such as quartz and rutile, in large amounts in the sediments we can speculate on the nature and intensity of the weathering climate. By contrast, the finer sediments of clay and silts suggest a much different climate scenario as well as that of deposition and stream dynamics.

Human changes of landscapes can alter runoff and the sedimentological character of fluvial systems. Dramatic changes in erosion due to the clearance of forests increase the sediment loads of streams causing aggradation and increased recurrence of floods. Since the Neolithic these changes have increased principally because of the introduction of plant and animal domestication. In extreme cases the soils of the watershed are eroded to bedrock, but more often the upper soil horizons (cf. Chap. 4) are decapitated leaving the resistant soil layers such as clay-rich B horizons.

Alluvial and colluvial fans: At Petit Chasseur, Sion, Switzerland during the construction of a new road between Sion and Petit-Chasseur in the 1960s, workers discovered the remains of over a dozen Late Neolithic (ca. 2500 B.C.) dolmen tombs (Osterwalder and Andrè 1980). Buried under deep (6 m+) alluvium, the elaborate tombs and 30 menhirs were well-preserved along a small tributary of the Rhône (Sauter 1976). During the construction and reuse by later peoples of the tombs, the Late Neolithic and Early Bronze Age climate produced nominal runoff and erosion. Sometime after 2200 B.C. and into the Middle Bronze Age, the climate became much wetter. Elevated levels in the Alpine lakes, which submerged the lake dwellings of earlier cultures attest to this. At Petit-Chasseur the increased precipitation, together with denudation of the surrounding slopes, led to greater erosion and runoff which led to increased flooding by the small river. Coupled with the aggradation of the river, colluvial debris flows inundated the site as well. The archaeological site disappeared under the alluvial fan that resulted.

2.4.2
Coastal Landforms

The fluvial delta marks the entry of the river to the sea. The delta is the alluvial fan deposited in the sea rather than in the stream valley. Coastal landforms are among the most changeable of all geomorphic settings. This is due to sea level change wherein a rising sea level creates a submergent coast and conversely, a falling sea level produces an emergent coast. In the US the bulk of the eastern shoreline is submergent while that of the west is largely emergent. In the Gulf of Mexico the shoreline of eastern Texas Louisiana, Alabama and Mississippi are submergent due the great sediment loading of the continental shelf by the Mississippi River. Coastal down warping, due to sediment loads, mimics that due to little or no tectonic activity producing continental margins like the eastern US. Coastal landforms can challenge fluvial for the density of archaeological sites. Wherein the fluvial system provided humans with rich riparian zones many coasts have comparably biologically active estuaries. Human cultures have taken advantage of these estuarine zones and their settlement remains dot the shorelines of the earth – past and present.

The most prevalent coastal landform is the beach, whether on a barrier island or simply the margin, above sea level, of a continental landmass. The beach is a transient topography changing over time with eustatic sea level and over the year according to storms and season. Ancient beaches are of archaeological interest. Henry de Lumley, in his landmark excavation of Terra Amata, demonstrated the presence of Pleistocene hominids – *Homo erectus* – on now uplifted, exposed beaches at Nice, France (De Lumley 1969). Beaches and dune systems often found at the back of them, provide direct clues to sea level and climate, fauna and flora found within their sediments.

Estuaries form at the back of the beach and represent the juncture of the river and the ocean. They are unique landforms that are the richest, in terms of biomass, along shorelines (Odum 1995). Estuaries are transient landforms like beaches. Unlike the beach the estuaries can penetrate inland and cover large areas of the coast. These landforms are rich in archaeological sites from the Mesolithic to the historic era. The deltas of large fluvial systems attracted early human groups to their rich alluvial deposits as well as abundant marine resources.

2.4.3
Glacial Landforms

When Louis Agassiz established glacial theory for good and all with his 1840 volume, *Etudes sur les Glaciers*, archaeology, as well as geology, adopted a powerful explanatory tool for landforms over much of the northern hemisphere (Agassiz 1940, 1967). Geomorphic evidence of former ice ages are often dramatic notably erosional features such as sculpted mountain terrains of cirques, arêtes, horns and hanging valleys. Depositional features, such as moraines, while less picturesque, mark the limits of advances and regressions of the ice sheets. In special cases like the Salisbury Plain, the great stones of Avebury and Stonehenge, likely originated as erratics carried by the movement of the Finnoscandic ice sheet.

The development of glacial theory, in the mid-nineteenth century, paralleled that of uniformitarianism theory and that of the antiquity of the Earth. Its impact on archaeological thinking then as well and now cannot be underestimated. What every school child learns as the "Ice Age", was not patently obvious to the geologists of the past century. Prior to the development of glacial theory, glacial features – moraines, outwash deposits, kettle lakes, erratics – were simply the result of catastrophic floods a.k.a. Noah's. Lyell was more realistic about fluvial processes and suggested that during the Noachian flood icebergs dispersed rock and sediment across the Earth's surface. If this had been true, Lyell should have wondered why the bulk of these glacial landscapes only occurred in the northern hemisphere.

Alpine farmers and mountain guides knew what Lyell did not, by way of daily experience. Glaciers even in their reduced modern forms advance and retreat, subject to climatic variation, pushing rocks, sediments and trees before them and then leaving this as debris when they subsequently melt. Ignatz Yenetz, a Swiss civil engineer, listened closely to a mountain guide, Jean-Pierre Perraudin, and in 1821, published a paper on the theory of glacial transport of exotic or erratic boulders (also called "foundlings" beyond the existing margins of Alpine glaciers (Hsü 1995). In 1829, he went further to propose an Ice Age wherein glaciers covered nearly all of Switzerland as well as most of Europe. Most geologists doubted such a theory with the exception of Jean de Charpenthier and his skeptical protégé, Louis Agassiz, Professor of Geology at the University of Neuchâtel. Because the scientific method works best when used to falsify a hypothesis or theory, Charpenthier and Agassiz set out to gather data to do just that. Their work did just the opposite. In 1840, Agassiz published his seminal work on glacial theory (Agassiz 1840).

Implicit in Agassiz's newly developed theory was the variability of the Earth's climate – cold periods led to glacial advances, warm periods led to their retreat. The question of what led to these cold periods or their duration was only to be answered in the mid-twentieth century by the general sequence of glaciations which was established in Europe and America at the start of the twentieth century by Albrecht Penck in Germany and T.C. Chamberlin in the midwestern USA (Penck and Brückner 1909). While presented as confirmation of the uniformitarianism principle (Tarbuck and Lutgens 1987), it was at that time perceived as anything but. Lyell's postulated uniformity of the Earth's processes from the beginning to "today" (Agassiz 1840) had no room for glacial theory (Hsü 1995). To make room for recent (Pleistocene) glaciations, Agassiz, with the help of Lyell's mentor William Buckland, took Lyell to glacial moraines within two miles of his home. Glacial theory, and "Ice Ages", was accepted by the British and Agassiz carried his message to America, visiting Harvard for some time.

Glacial theory has an immediacy for archaeology that stems from its description of an Earth, in the most important period of humanity's development, the Pleistocene. For the European archaeologists no reading of fossil sites or material culture could be made without an eye toward how they fit into the ebb and flow of glaciations. For Americanists, the great continent-wide ice sheets presented a barrier to colonization of North and South America. Human settlement of the Americas, by migration from Asia, had to wait for the right convergence of natural opportunity and human necessity. This is generally believed, to have been during the terminal Pleistocene sometime after 20,000 years ago.

The most dramatic archaeological discovery in the last decade of the twentieth century came at the Tyrolean glacier, on the Alpine border of Austria and Italy, with the discovery of the now-famed "Iceman" mummy (Spindler 1994). Found at an altitude of 3000 m, this intact prehistoric man and his accoutrements, reminded archaeology of the importance of glacial landforms to human society as well as their potential for study. The sacrificial mummies of the high Andes, at even higher altitudes – 6100 m – are a testament to the importance of alpine regions to ancient cultures (Carey 1999).

It is their linkage to climate that makes the glaciers and landforms derived from them important to human prehistory. Like the volcano and river, the glacier has shaped landscapes, started new soils and, therefore, created new ecologies over 30% of the Earth's land surface. The great loess (fine silts) deposits of periglacial Europe, America and Asia provide evidence of ancient wind fields and paleo-circulation patterns. Riverine dunes and other inland dune systems were active in glacial periods while today they form substrates for riparian forests or expansive grasslands. Meltoffs of the ice sheets dramatically shaped drainage systems far beyond the ice margins. Transport and deposition of sub- and proglacial sediments by outwash streams along flood plains and terraces provided attractive locations for colonization by early farming cultures. Moraines and till plains were commonly exploited by early cultures for lithic materials.

Landscapes and glaciers. The Rhoads Site, Illinois illustrates the influence of the North American glaciers on a prehistoric site in central Illinois (Follmer 1985). The site is located on a till plain of the Illinoian (Pentultimate) glaciation and lays just 14 km south of the Wisconsin glaciation's terminal moraine. This unique location allowed an examination of the surface geology and soils that had come under the direct influence of two ice sheets. Between the retreat of the Illinoian ice sheet around 130,000 years ago, and the return of the ice with the Wisconsin the interglacial period called the Sangamon in North America, soils formed on the Illinoian till. These soils are quite mature and deep – ca. 3 m. The early Wisconsin advances did not intrude as far south as Illinois so the soils remained undisturbed until 25,000 years ago with the full glacial maximum. At lower elevations along a small drainage near the site, outwash deposits began filling the stream valley while loess (Peoria) accumulated on the uplands of the site. Both in-filling and loess deposits stop at 12,600 years ago . Follmer readily identified two soil types – mollisol and alfisol – in the area of the archaeological site. The former soil type is formed in grasslands while the latter type forms under deciduous forests. Subsequent Wisconsin glaciation deposits – alluvial and aeolian – buried both the mollisols and alfisols of the Sangamon Interglacial. Where not so deeply buried or exhumed, mollisols of the interglacial saw formation of later alfisols. As a rule the stratigraphy indicates a succession of buried paleosols and post-Wisconsin soils. When human populations colonized the area, Follmer identifies a mosaic of prairie and forest based on the patchwork of soils on the Wisconsin till (Follmer 1985). We shall continue this example later in Chapter 5 as illustrative of key descriptive soil parameters.

Interestingly, it is during the Pentultimate glaciation (ca. 180–130,000 years ago), and the last ice age (ca. 75–20,000 years ago), that we see the colonization and proliferation of *sapiens* cultures in the northern hemisphere. Both Neanderthal and later populations lived in close proximity to the ice sheets preying on the megafauna of that time. For the New World, the debate of colonization and the influence of the ice sheets has intensified. Like the Old World, these early New World cultures were Paleolithic.

2.4.4
Desert/Arid Landforms

Just as near shore continental shelves were once coastal plains, dry lands and deserts were often verdant savannas and forests. Drylands form 35% of the earth's land surface (Oviatt et al. 1997).Within these drylands are true deserts (16%). Water plays an important role in shaping dryland topography, but a more significant factor is the wind. The reason the wind is important is self-evident from an inspection of the term "dry" land.

Dryland landforms created by water vary from arroyos ("wadis" in Old World parlance) to eroded mesas and buttes. Playas form in deserts where precipitation is insufficient to maintain permanent lakes. Where lakes do exist in dryland areas such as the Great Salt Lake (USA), the Dead Sea (Israel/Jordan), and Pyramid Lake (USA), snow melt and other runoff feeds the basin from outside the immediate area of the lake itself. Ancient fluvial systems or paleo-drainages have been revealed in the shallow subsurface of the eastern Sahara in the driest region of southern Egypt and northern Sudan (McCauley et al. 1982). Space Shuttle-based radar provided us with evidence of a long and complex fluvial history in a now hyperarid and uninhabited region (McCauley et al. 1982). Well-defined networks of broad alluvial valleys, braided streams together with bedrock-controlled channels appear in the radar images and have been confirmed on the ground by archaeologists. Along these "radar rivers" have been found evidence of some of the earliest inhabitants of northeast Africa and northwest China (Wendorf, in press; Holcomb 1996). Allogenic streams exist in many arid lands – the Nile in Ethiopia and Egypt, the Jordan in the valley and hills of the Dead Sea, the Truckee ending in the Nevadan desert at Pyramid Lake. and the Colorado of Grand Canyon to the Gulf of California. These fluvial systems can be sensitive indicators of paleoclimate because of their aggradation and erosion due to rainfall outside the bulk of their drainages (Hassan 1988).

In the midwestern United States are landforms that dramatically display the interplay of wind-born deposition of both sand and silt (loess). These are the sand hill regions of the state of Nebraska along with the loess-prairies of Nebraska, Kansas and the Missouri River valley (Fig. 2.1a,b). Most identifiable of aeolian landform are dunes – barchan, transverse, parabolic, longitudinal and linear (Fig. 2.2). When active these features were not likely to have been the locus of much human settlement, but after the establishment of a vegetation cover, they have been occupied by both pastoralists and agriculturalists. Geoarchaeological studies of artifacts and buried surfaces in aeolian deposits can provide significant information for the archaeologist (Leigh 1998).

2.4.5
Karst/Cave Landforms

Nowhere is the linkage of hydrology and geology so apparent than in these types of landforms and features. In some cases the situation is one of underground drainages carving vast solution features rather than great fluvial valleys. Karst landforms are generally considered to originate on carbonate rocks, in most cases, limestones. This limestone can be composed of 50% or

[handwritten margin note, left side:] ALLOGENIC: (OF A MINERAL OR SEDIMENT) TRANSPORTED TO ITS PRESENT POSITION FROM ELSEWHERE.

[handwritten margin note, left:] mP

[handwritten note, bottom:] KARST : LANDSCAPE UNDERLAIN BY LIMESTONE THAT HAS BEEN ERODED BY DISSOLUTION, PRODUCING RIDGES, TOWERS, FISSURES, SINK HOLES, AND OTHER CHARACTERISTIC LANDFORM

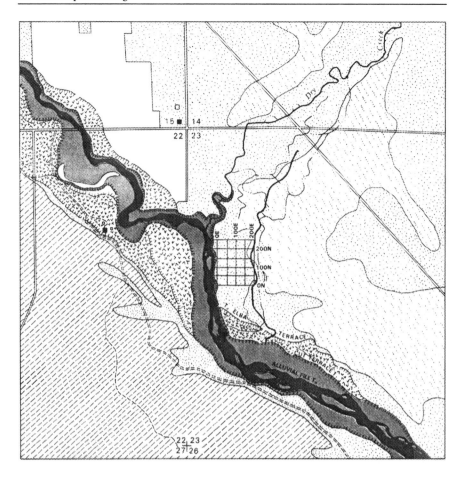

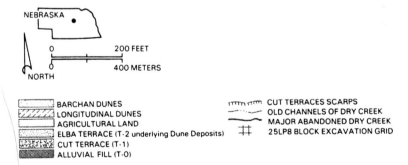

Fig. 2.1. a Geomorphic map of a reach of the Loup River, Nebraska. Aeolian and fluvial geomorphic features are highlighted. *Grid* in center denotes archaeological locale (adapted from Roper 1990). b Region map for Sand Hills and Loess Plains of central United States. (Modified from R.M. Busch, ed., *Laboratory Manual in Physical Geology*, Fig. 11.2, 1996)

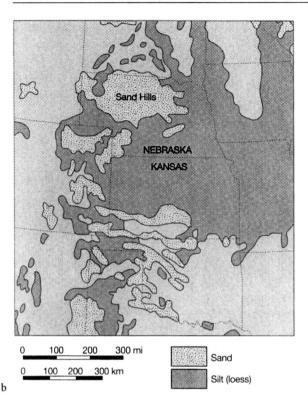

Fig. 2.1. *Continued*

0   100   200   300 mi

0   100   200   300 km

Sand

Silt (loess)

b

more calcite or aragonite, both pseudomorphs of CaCO$_3$. In the Ozark Mountains of the central US karst forms in dolomites – CaMgCO$_3$ – that is more than 50% of this facies, but more often limestone is the rock system (Easterbrook 1998:194). Springs are very common in these landforms. In Missouri, heartland of the Ozark Mountains, the dolomites are rich in caves and springs (Vineyard and Feder 1982). In the gold reefs of South Africa's Witwatersrand the massive dolomite strata contain vast quantities of water that are of concern to miners (Garrison 1999).

Precipitation is critical to the production of karst, but only if the water table is high enough to facilitate the dissolution process. Therefore development of karst is, like the fluvial landform, linked directly to climate and, importantly, the geological structure (Palmer 1991). Faults, joints and fractures form the conduits for water and entrained chemicals, such as carbonic acid that dissolve the limestone. Where sulfide deposits exist along fault planes the production of sulfuric acid will accelerate the growth of karst.

MP The most recognized karst landform is the sinkhole or doline. These closed depressions are solution/collapse features that vary around a basic, circular or elliptical, funnel-like shape. In some of these the local hydrology forms a

FACIES: THE CHARACTER OF A ROCK EXPRESSED BY ITS FORMATION, COMPOSITION, AND FOSSIL CONTENT.

Fig. 2.2. Dune forms:
a parabolic, **b** barchan,
c transverse, **d** longitudinal,
e linear. (Modified from Conte
et al., *Earth Science*, Wm. C.
Brown, Fig. 12.28, 1997)

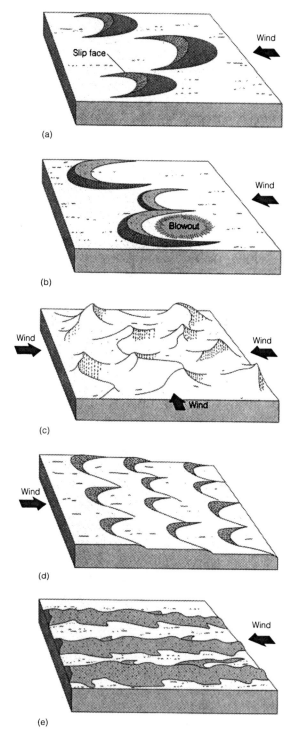

spring. For archaeology these sources of water have concentrated both fauna and human cultures forming deposits of both within and around the spring. Where the water has abandoned the subterranean feeder conduits of the ancient springs, caverns and caves now exist. Collapses occur over time that form *traps* for animals and humans. In Spain, collapsed limestone caves have produced some of the most spectacular evidence of early humans on the continent of Europe. The collection of sites known as Atapuerca lie within the northern sierra of Spain. The caves, known since 1976, have since the beginning of serious excavations in the 1990s, produced over 1600 human bones ranging in age from greater than 780,000 to 127,000 years ago. This chronology has been derived from work in what is termed "the Gran Dolina" or Big Sinkhole (Bahn and Renfrew 1996).

The nearby site Sima de los Huesos or "Pit of Bones" has yielded the bulk of the large collection of human remains found to date and revolutionized our thinking of the colonization of Europe by early humans. Indeed, in Europe and elsewhere, some of our earliest fossil humans and their material culture have originated in caves. The first of these was Neanderthal, discovered in 1856 in its eponymous valley. A continent away, *Homo erectus* known as "Peking Man", was found in the debris of the "Cave of the Dragons" – Zhoukoutien.

In 1990 and shortly thereafter in 1994, two dramatic cave art discoveries were made in France. These solution features in massive limestones, south of the Massif Central, were respectively, Cosquer Cave and Chauvet Cave. Both contain some of the oldest Upper Paleolithic art ever discovered – 29,000–35,000 years (Clottes and Courtin 1996; Clottes 2001). Both are located outside the heartland of Paleolithic cave art, the Loire River valley and northern Spain. Cosquer Cave is unique, by virtue of its discovery below the surface of the Mediterranean near Marseilles. The cave entrance was found 30 m below the sea surface by scuba divers. Its entry, by ancient artists, occurred sometime during the last glaciation (Würm/Weichsel). Chauvet Cave, located along the Ardèche River, a tributary of the Rhône, lies on the south flank of the Massif Central and dates to the same period as Cosquer.

2.4.6
Volcanic Landforms

Not every geomorphologist considers this type of landform within the scope of basic geomorphic settings. This is somewhat surprising given the prevalence of volcanism and its impacts on humanity over time. Islands are almost always volcanic landforms. Continental terrains are imprinted with the remains of volcanic events. The history of archaeology would have been much different were it not for the discovery of the lost Roman cities of Pompeii and Herculaneum in the later 1700s (Corti 1951). In the modern world we are reminded almost annually of the force of volcanoes and their disruptive impacts on nearby human groups. At the global scale we can consider volcanism a factor in shaping climate (Bryson 1988).

Volcanic landforms range from the signature cone shape to craters, caldera, tabular mountains, basalt flows, ashbeds to exposures of intrusions of magma such as plutons, dikes, batholiths, sills, cuestas and pinnacles. There are flood plains called llanos created by lava dams of drainages (Inbar et al. 1994). Volcanic landform can, like the volcano itself, be categorized as active, dormant or extinct. Iceland is both an island and a ongoing display of a variety of volcanic landform. In western Europe past eruptions of Iceland's volcanoes have acted as "clocks" with widespread tephra deposits blown from these sources. Geoarchaeologists can map these deposits and gain an understanding of the timing of human occupation at a location far remote from the eruption site. Deposition processes associated with volcanoes are among the most dramatic in nature. After deposition these volcanic deposits are exposed to erosional processes common to other geomorphic settings.

The following statement from Luciana and Tiziano Mannoni (1984) provides adequate justification for archaeological geology's interest in volcanoes and tectonic processes: "The movement of the plates and continental platform within the covering crust causes volcanic eruptions and earthquakes. Crustal subduction, orogenesis, and the metamorphism associated with crystal deformation, under great pressure and temperature, at plate boundaries create earthquakes and the formation of volcanoes. All of this is subsumed under the rubric of the Wilsonian Cycle of plate displacements'" (Wilson 1966; Hsü 1995). One has to go no further than the names of Pompeii, Troy, Thera, Port Royal, Jericho and Ceren to assign landmark archaeological/geological significance to cataclysmic eruptions and earthquakes. Pompeii and its neighbor Herculaneum has the place of honor in the pantheon of archaeologically (and historically) documented catastrophes.

From August 24 to August 25, 79 A.D. Vesuvius erupted in what has been named for its chronicler, Pliny the Younger, a Plinian eruption. Hal Sigurdsson (1999) published a detailed discussion of Pompeii's end. From an archaeological standpoint, Pompeii's discovery in 1748 led at first to unsystematic antiquities collecting, but by 1860 and the appointment of Giuseppe Fiorelli, well-documented excavations began and have continued to the present. Tourists including Lyell came to Pompeii both as antiquarians, and in Lyell's case, as a geologist. Sigurdsson uses the Latin texts and his own analysis of the tephra deposits, in a chronological study of the fatal eruptive sequence. Herculaneum was destroyed on the first of the 2-day eruption.

Thera, the volcanic island also known as Santorini, buried the Minoan town of Akrotiri. Like Pompeii, Akrotiri was buried under several meters of volcanic ash. It was discovered in 1967. The eruption of Thera occurred along the Hellenic Island arc, a back-arc volcanic region formed in Miocene-Pliocene times (Hsü 1995). Wrapped up in the eruption of Thera, a Krakatoan-type blast that has also been alternately blamed for the eclipse of Minoan civilization as well as being the inspiration for the Atlantis myth. As to the former charge, Thera has been found not guilty by both archaeology and earth science. Spyrudon Marinatos, first excavator of Akrotiri, speculated on the timing of the Theran eruption being coincident with the archaeologically observed destruction of Minoan Palace culture on Crete 100 km to the south of Thera. Placing the eruption at around the time of the Late Minoan 1B period (ca. 1450 B.C.) helped solve questions of the demise of high Minoan culture, but posed other problems for later archaeologists at Akrotiri. No Late Minoan 1B ceramics have been found there, only earlier Late Minoan 1A period wares (ca. 1500 B.C.). Based on this discrepancy, many archaeologists see no direct connection between the Thera eruption and Cretan problems caused by the tsunamis and ash falls. The recent publication of geochemical data from the Greenland ice sheet cores (Fiedel et al. 1995) indicate a sharp increase in volcanically derived concentrations of NO_3^- and SO_4^{2-} in the atmosphere in the interval between 1646 to 1642 B.C. Coupled with dendrochronological data for the same period, 1628–1626 B.C., a major volcanic eruption occurred, but was it Thera? The evidence argues against a fifteenth century B.C. eruption and suggests early mid-seventeenth century B.C. for the Theran eruption.

2.5
Earthquakes – Volcanic or Otherwise

Earthquakes, whether associated with volcanic eruptions or, the result of fault movements, alter landscapes – natural and cultural. Amos Nur, Stanford geophysicist, has investigated earthquakes in the Holy Land (Nur and Ron 1997). Earthquakes are the second consequence of tectonism – volcanoes being the other. Indeed, where there is volcanism so there will be earthquakes, but the opposite is not necessarily so. In the collision of continental plates, volcanism or magmatic arcs are not present or occur only early on in the process. The collision need not be "head on", a convergent boundary, but a "side-swipe" type of collision, called a transform fault boundary. The boundary between the Pacific and North American plates is a transform boundary as is that between the Mediterranean and Arabian Plates. Both are the seismically active zones, particularly in California and the Jordan-Dead Sea Transform fault. Along the Dead Sea–Sea of Galilee axis, Nur identified a 14-m strike-slip movement in the past 2000 years. The observed motion of the fault accounts for only 3–4 m of this amount. Why the discrepancy? Nur suggests the occurrence of magnitude 8.0 or greater (Richter scale) event(s) in the past.

Using literary sources, histories and the Bible, Nur has accounts of major earthquakes in 370 A.D., 31 B.C., and 756 B.C. Judea, half of the dual monarchy of Israel and later the name of the Roman province, experienced a devastating earthquake mentioned in Josephus at the same time as the Battle of Actium (31 B.C.). The 370 A.D. quake destroyed the Roman city of Beth Shean and is mentioned by Ammianus Marcellinus (Nur and Ron 1997, 55). This event was recorded in Cyprus, 360 km to the southeast. Zachariah describes a major earthquake, along with the classic strike-and-dip movement, at Jerusalem in 756 B.C. The strongest candidate for earthquake destruction is ancient Jericho where "the walls come tumbling down". Jericho is no stranger to earthquakes, being struck in 11 July 1927 by a major event that destroyed major portions of Jerusalem killing hundreds.

2.6
Mapping Techniques

Identification of particular landform is typically done by use of (1) planimetric maps, (2) aerial or satellite imagery and (3) ground reconnaissance. The objective is to map the principal geomorphic landform from remotely sensed data onto maps of scales appropriate to the archaeological questions being asked. Broad geomorphic features, at local and regional scales, can be deduced from aerial/satellite images. In regions with extensive vegetative cover conventional visible-spectrum photographs are not as effective for evaluating landforms as those using infrared and microwave regions of the electromagnetic spectrum. These latter types of imaging can aid in landform classification by differing levels and types of vegetative cover, surface hydrology and, using newer types of high-resolution imaging spectrometers, mineralogy and rock types can be examined. Ground level studies include walkover surveys and subsurface studies using both hand and mechanical means.

From these images and ground investigations the geomorphic map should describe: (1) morphology (2) landform (3) drainage patterns (4) surficial deposits and (5) tectonic features together with geomorphic/geologic processes active in the mapped area. In today's world this is where the value of GIS is most apparent. Once the geographic information is in digital form GIS programs can be used to make a base geomorphic map (the *primary surface* in GIS parlance). Then a series of subsequent maps (*secondary surfaces*) showing elevation data, stream locations, soil types, landform types and landform units such as alluvial fans, colluvial deposits, terraces, etc. By breaking up the geomorphic data into multiple displays one can avoid the problem of complicated and difficult-to-interpret maps. In the US and many other developed countries it is possible to obtain geographic and geologic data encoded into computer-compatible form. In the US primary source would the US Geological Survey (USGS). Digital terrain maps are available from that agency while coastal maps are available from the National Oceanic and Atmospheric Administration (NOAA). Digitized topographic maps now exist at a scale commonly used in archaeology – 1:24,000 – that of the 7.5-min quadrangle map. Many countries have similar maps available. The GIS base maps of Mexico, for example, are exemplary.

Primary data are obtained from a variety of sources from traditional maps to satellite images. If one starts with satellite imagery such as LANDSAT3 then the scales of this format vary from 1:250,000 to 1:1,000,000. At the larger scale the images are 185 km on a side with data on several spectral bands that are available upon request. The most commonly used bands for geomorphic mapping are the visible and infrared bands 4–7 with composites available. Band 4 is the green spectral band and is good for delineating areas of water in coastal areas; band 5 is the red band and is good for the vegetation mapping; band 6 is the thermal infrared band which discriminates between land and water; band 7 is the near-infrared spectral band which penetrates haze emphasizing water and boundaries in landform. The LANDSAT Thematic Mapper (TM) provides 7 bands of spectral data with an image resolution or pixel size of 30 m. This is an improvement on original LANDSAT resolution which was 80 m. New commercial satellite imaging by the IKONOS satellite, launched in 1999, produces 1 m pixel visual band images. The cost is about $1000/negative. This availability of high resolution imagery promises to be a boon to archaeo-geomorphic mapping efforts.

The high resolution infrared spectrometer (HRIS) has a continuous sampling range from 0.4 to 2.5 µm region with a capability of 10-nm spectral sampling intervals, thus obviating the "band" restrictions of lower spatial resolution systems such as LANDSAT MSS and TM as well as the French SPOT systems. The continuous spectral capability of the HRIS gives the researcher the ability to assess sediments, rock types and minerals in exposed terrain (EOS 1987:2–3). For example the spectral reflectance of sedimentary rocks, which constitute about 75% of the Earth's surface (Pettijohn 1975), varies to a degree that important facies such as calcite, which absorbs in the short wavelength infrared, can be readily identified along with limestones, shales and sandstones. Using LANDSAT TM bands 4, 5 and 7, Carr and Turner (1996) have demonstrated different spectral profiles for rock, soil and weathered bedrock in western Montana. Figure 2.3 shows HRIS spectra of altered rocks. For relief, elevation and penetration of vegetation, radar – side-looking airborne radar (SLAR) and synthetic aperture radar (SAR) – in the C- and L-bands have proven useful (Gibbons 1991) as the Egyptian and Chinese results have shown (McCauley et al. 1982; Holcomb 1996; Wendorf in press).

Both aerial photographs and multispectral data are commonly had at scales of 1:80,000 down to 1:2000. In practice these can be co-registered with satellite data at comparable scales or simply used singularly. (Fig. 2.4) Aerial images are of three types – vertical, oblique and mosaic. The most commonly used are the vertical images which form the mosaic type when co-registered with adjacent images. Both are used in geomorphic mapping depending upon the scale of interest. Oblique aerial photographs provide a wide-angle, perspective view of terrain and assist in determining low-relief features in relatively flat terrain. In rough terrain the value of oblique images is reduced. For the detection of archaeological features the aerial image is still unsurpassed.

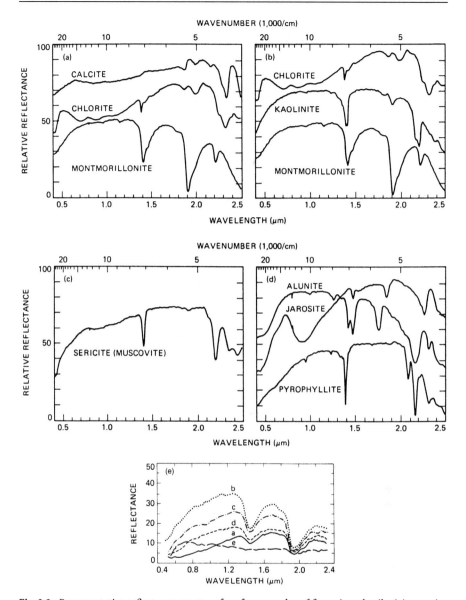

Fig. 2.3. Representative reflectance spectra of surface samples of five mineral soils: (a) organic-dominated (high organic content, moderate fine texture); (b) minimally altered (low organic, low iron content); (c) iron-affected (low organic content, medium iron content); (d) organic-affected (high organic content, moderate coarse texture); and (e) iron-dominated (high iron content, fine texture).

Fig. 2.4. Co-registration of four remote-sensing formats – LANDSAT, SIR/SAR, SPOT and aerial photograph. (Modified from Comer, *CRM*, no. 5, p. 11, 1998)

Again this is due to the scale of interest. Features of geomorphic interest are not necessarily those of archaeological interest and this must be constantly kept in mind when selecting remote sensing data for mapping studies. Archaeological sites themselves are rarely detected from satellite images (McMorrow 1995) so *site* scale studies should rely on the advantages of aerial images of lower scale, higher resolution.

Some cautionary notes on using remote sensing imagery as a substitute for planimetric maps should be made here. When using these media either to create a geomorphic/geologic map or augment a planimetric map, there are inherent problems both in spatial perspective and elevation. If an image has not been acquired at the truly vertical then the image lacks spatial fidelity across its height and breadth. The only point on the image of spatial accuracy is directly at the *nadir* or point directly below the lens. All other areas of the image are spatially distorted with respect to their true positions. That is to say the ground distance between geomorphic features is not in a true proportion with the photo distance between those features no matter what scale is used. In addition, the heights of features are distorted such that relief is exaggerated except in the flattest of terrains.

When photomosaics are made from a collection of vertical images the same problems still ensue. Only the area around the nadir or principal point will have reliable spatial relationships. Those toward the edges of the images are the most distorted. This distortion can be removed by the process of *rectification* where all points on the image plane are corrected to the vertical. Vertical exaggeration is harder to remove and depends a great deal on the focal length of the lens making the image. Longer focal length lenses produce less vertical exaggeration as a rule. When image distortion – both spatial and vertical – has been removed the image is known as an *orthophotograph or orthoimage*. Mosaics made of these are truly picture maps and can have contour lines overprinted on them. When these mosaics are overlain on planimetric maps the best of both worlds is achieved in the creation of the *orthophotoquad* map.

2.7
Scale

Generally, geomorphology, in the service of archaeology, is not practised at the global scale. We are principally concerned with topography and its control of surface hydrology and past human ecosystems. Scale is important in the conduct of geomorphic and geologic mapping. In the remote sensing of topography, from aerial- and space-based platforms, the National Aeronautical and Space Administration (NASA) has recognized three scales of interest: global scale – 1 km horizontal, 10-m vertical resolution; intermediate scale – 100-m horizontal, 1-m vertical resolution and local scale – 10-m horizontal, 0.1-m vertical resolution. These scales apply to both visual and multispectral imaging.

Their usefulness in geomorphic mapping is fairly obvious. When mapping we refer to high resolution as small scale. In terms of map scale this nomenclature is confusing to many. "Large-scale" maps of 1 : 150,000 to over 1 : 1,000,000 are "low resolution". High resolution at the local scale is represented by ratios of 1 : 20,000 or smaller. Observations made at different scales are useful depending, of course, on the types of questions being asked. One must be careful in geoarchaeology (Stein 1993), to address scale from the human perspective. In archaeology landscape reconstructions must be focused on their relevance to human cultures, their processes and consequent stability or change over time. In this regard geoarchaeology is differentiated from traditional geomorphology and other geosciences by its use of earth science methods to make interpretations about human groups interacting with landscape, climate, vegetation and hydrology (Stein 1993).

Mapping archaeology onto topography is done more often at local or intermediate scales. Almost all topographic questions are related to inter-site mapping, but with significant exceptions for large urban sites such as Great Zimbabwe, Cahokia, Teotihuacan, to name some of the more obvious examples which benefit from mapping at the intra-site level.

2.8
Making the Map – One Example

In the digital age a geomorphic map or geologic map can be produced with media downloaded from cyberspace. On a recent field project my colleagues and I obtained digital photographic images of the area of the archaeological site where we were working. These we attempted to reconcile to planimetric maps of that same area. Our first concern was that of *scale*. The two formats differed significantly such that they had to be *ratioed* where the two formats were in a realistic numerical proportion. Another concern was when the maps and the images were made, e.g., the *date* of their production. Photographs and other imagery differ significantly from maps in their dates. A map is a synthetic creation being a generalization of the temporal frame which is less important than that of the spatial. In particular, if a map has vegetation printed on it then its distribution is that of the time period when the map was produced. Compared to a satellite image made 10 years later there may be dramatic differences. In our Greek example, discussed in the next section, the area of our site is intensively cultivated so the vegetation pattern changes quite significantly with each planting season.

While less changeable, one must consider the hydrology of an area. In our case the small drainage adjacent to our site was shown in basically the same location and pattern on both map and photographs. In active fluvial systems one may not be so lucky. Channels shift laterally, islands can aggrade or erode and whole terrace systems can be modified by large floods. Located next to a small ephemeral stream and flanked by two rocky hills, the geomorphic situation of our site was hardly complex. These fixed features gave us good reference points to make our geomorphic map. Working with digitally based media allowed us to utilize another recent development of the space age – global positioning satellites or GPS. Using hand-held receivers one can quickly map the geomorphic and geologic features. An example of a geomorphic mapping schedule that can be used follows:

1. Review aerial photographs of the area of interest. Use rectified/ratioed copies where available.
2. Review LANDSAT or appropriate remotely sensed data of the area of interest.
3. Visit the field site. Survey when vegetation is less of a problem (fall, winter).
4. Develop a landform classification scheme of the area – terraces, dunes, mesas, etc.
5. Select the scale of the geomorphic map or in GIS – the primary surface (1 : 80,000, 1 : 24,000, etc.).
6. Classify landform from the imagery using quantitative or qualitative parameters – relief, elevation, gray-scale, color, pixel-size, spectral reflectance, etc.

GEOMORPHIC Mapping Schedule

7. Transfer the classifications on to the preliminary geomorphic map.
8. Ground truth or field verification, by walkover survey and subsurface excavation, the geomorphic classification of the geomorphic map. This is particularly important in highly vegetated and forested terrain.
9. Finalize the geomorphic map – scale, legend, colors, projection, etc.

2.9
Lithares, Boeotia (Greece) – Geographic Information Systems, Geophysics and LANDSAT

Lithares is an Early Bronze Age (Early HelladicI/II) settlement in Boeotia near modern-day Thebes (Figs. 2.5 and 2.6). It was surveyed and excavated in the 1970s (Tzavella-Evjen and Spyropoulos 1973; Tzavella-Evjen 1985). In 1999 Professor Tzavella-Evjen returned to Lithares with a team of archaeological geologists from the University of Georgia and the US Department of Agriculture's Forest Service. A week of shallow geophysical survey and GPS mapping were carried out at the site. The results are shown overlain on aerial photographs and GIS maps of the site locale. As indicated above, the first steps in our study of Lithares was to acquire both aerial and satellite imagery together with geological maps. While not slavishly following our protocol, Figs. 2.5 and 2.6 illustrate a procession from large scale to smaller scale maps that locate Lithares in the overall region – ca. 1 : 3,000,000 scale down to area geology (Fig. 2.5) and finally to mapping local scale site features – drainages, land use, geomorphology, etc. – onto aerial imagery (Fig. 2.6).

The 1999 study was able to merge data from field and photo-geology done in 1964–1965. Geological mapping by the Greek Institute for Geology and Subsurface Research was not complete at the time of the earlier studies of Lithares. In addition, smaller scale geological maps of Boeotia and the Theban Plain – 1 : 200,000 or less – have allowed Higgins and Higgins (1996) to sum-marize the structural geology in regards to archaeological interests. In our study we were able to add to the synthesis by Higgins and Higgins (1996) by recourse to field mapping of the geology and geomorphology at Lithares *per se.* For instance, Higgins and Higgins show the local geology of the south shore of Lake Iliki, the municipal water supply for modern Athens, as that of a Triassic – Jurassic limestone (Jt). Our field studies and inspection of recent geological maps of Boeotia and nearby Euboea indicated a greater diversity in the local lithology which, in turn, does much to influence the topography and geomorphology at Lithares.

The regional geological setting is well described for central Boeotia. The Theban Plain or Plains are graben (a fault-bounded crustal block, lower, relative to the surrounding blocks) within the surrounding limestone hills (Higgins and Higgins 1996). Lithares itself is located in a low saddle and valley between two small hills (~350 m) on the east and west of the site while being bordered on the north by the lake (Fig. 2.5). A small drainage along the east

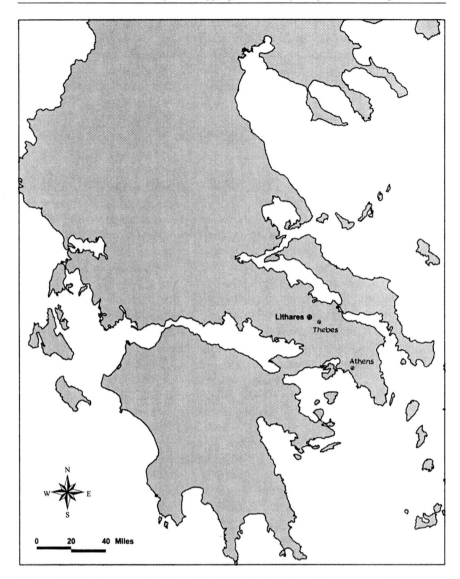

Fig. 2.5. Site location map for Lithares Neolithic site, Boeotia, scale 1:3,000,000. (Courtesy of Glen Lacascio)

side of Lithares empties into Lake Iliki (Fig. 2.6). The soils ("agrosols") of the saddle and valley are intensively farmed and are quite deep (>2 m) of grayish, fine-grained alluvium-colluvium. A characteristic of EH I/II settlements (2300–1600 B.C.), is the extensive use of stone in domestic and public architecture. The site is covered with the remnants of low walls made of limestone and other sedimentary facies. Given the local geology it is clear these

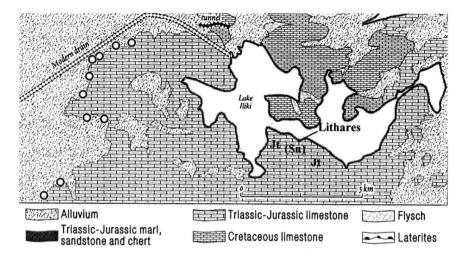

Alluvium

Triassic-Jurassic marl, sandstone and chert

Triassic-Jurassic limestone

Cretaceous limestone

Flysch

Laterites

Fig. 2.6. Geological map of Lithares locale. Modified from Higgins and Higgins (1996). *Inset* Detail of tilted limestones at Lithares. (Photograph by the author)

materials were quarried/collected nearby. The low hills adjacent to the site were the likely candidates for building stone extraction, given their unique tectonically upturned strata (Fig. 2.6).

Greek geological maps, unlike that of Higgins and Higgins, show Lithares and the south shore of Lake Iliki to be situated on a complex of schists and cherts, together with radiolarites, sandstones, and white limestones (Sn), whereas the deeper lithology is a more uniform dolomitic limestone. This latter identification, Sn, has been added to their geological map by the author based on field observations in 1999. The two hills are composed of the mixed lithology with few, if any outcrops, of the dolomitic facies seen. The colluvium formed from the weathering of the hills has been mixed with alluvium from the valley drainage as well as high stands of the adjacent lake in the past. The soils are quite stony in character although extremely fertile. With the exception of the lake little modern surface water is observed, although in antiquity the now-drained, large Lake Copais lay only a few kilometers to the northwest of Lithares. Today water for irrigation is taken from deeper aquifers within the limestones.

2.10
Geological Mapping

The mapping of geological sections and structures has been a traditional capstone technique taught to earth science students. By understanding the horizontal and vertical relationships of geological strata across space the extent and nature of depositional and erosional sequences are realized. The principal differences in geological and geomorphological mapping are the scale and the nature of what is being mapped. Scale is one thing while landscape and lithology are quite another. Both use the same basic protocols on different scales of lithostratigraphy. Both scales have relevance for archaeological inquiry.

For instance, the location of resources such as flint or marble will generally signal the likelihood of nearby habitation and/or extraction sites, e.g., Lithares. Conversely, the presence of these materials in the archaeological context requires the identification of the source location. Either way the mapping of the geology of an area is required.

2.11
The Vertical or Stratigraphic Section

Formations make up geological sections that in turn are composed of rock *strata* or *beds* organized into *members* which, unless they are of igneous origin such as basalt or granite, were laid down according to the Principle of Original Horizontality which presumes a depositional history of a uniform lenticular nature. Intrusive features – dikes, sills, etc. – are the result of volcanic processes which, in turn are the result of the tectonic activity due to the restless nature of the continental plates. *Structural* features within sections such as faults, synclines and anticlines, folds and their like are indicative of deep Earth seismicity as well. *Stratigraphic* relationships such as conformity and the lack thereof express the broad patterns of the rocks in the Earth's crust. A geologist reads the geologic section as a series of pages of the Earth's history with all the twists and turns of that record imprinted therein.

For the early part of humanity's own story, the rock record was the only reliable means of assessing that antiquity with any assuredness. Following the protocols of stratigraphy early antiquarians and later archaeologists were able to deduce precedence, succession and simultaneity of fossils and ancient tools found within the *members* of a formation. Necessarily, the primary relation was simply that the oldest strata were at the bottom of the section – Smith's Principle in its most simplistic expression of the *Law of Superposition*. As mentioned previously, Sir John Evans and Joseph Prestwich's visit to Boucher de Perthes' excavations in the Somme River terraces provided the geological framework for assigning the true antiquity to the latter's claims for the relationships between ancient fauna and tools within the gravels.

The relative age of a lithostratigraphic formation's bed or member (a bed being a subunit of a member) can be determined by superposition and the *Principle of Inclusions* which holds that inclusions or fragments in a rock unit are older than the surrounding unit itself. For instance, in the case of lava or tephra deposits, those inclusions, geological or cultural, are by definition older than the volcanic matrix. In the case of volcanic flows or ash falls, such as those of Pompeii and Ceren, cultural debris was entrained or enclosed in the deposits and provided what are termed a *terminus post quem* – that is the deposit cannot be older than the inclusions. Before the advent of chronometric dating techniques, archaeology relied on this principle of stratigraphy without specifically stating it. In East Africa the finds of early hominids have come from sedimentary sequences of ancient lake beds such as Olduvai gorge (Fig. 2.7).

Hans Reck, a member of the Geological Survey of German East Africa, and later Richard L. Hay (1976), untangled these sequences for the Leakeys – Mary and Louis – enabling the latter to more confidently assign the Pleistocene ages to finds such as *Zinjanthropus* (now classified as *Australopithecus*) and *Homo habilis* (Fagan 1994). The Olduvai sequence is a model of the use of stratigraphic principles with the straightforward terminology of Bed I, Bed II, Bed III and Bed IV. A Bed V, relatively thin and recent, was identified at the top of the Olduvai section being what is called the *end member* as well. Reck distinguished these beds within the Olduvai section based upon their color, sedimentological texture and *biostratigraphy*. The latter procedure relies on the identification of fossil forms – macro and micro – to help assign a *chronostratigraphic* or "chronozone" sequence to a section.

Lithostratigraphic units are defined by physical attributes of rocks where similarity in lithology implies a geographic and genetic association in the units. A lithostratigraphic correlation does not necessarily imply a temporal equivalence. At Olduvai, biozones based on fossil genera can provide a rough temporal control. "Rough", in a temporal sense, implies that the presence of a particular fossil type – *index fossil* – only indicates the time range of a plant or animal species – fine for geology, but not very useful for archaeology. At Olduvai biostratigraphic correlation was useful because of the millions of years that the strata represented.

Olduvai Beds I and II represent the most ancient lithified remains of sand and clay-rich lacustrine environments of a late Pliocene/early Pleistocene. Bed III represented a dramatic change in color compared to the lower two beds. It is strikingly red in color and contrasts sharply with its whiter stratigraphic neighbors. The next upper member was named Bed IV.

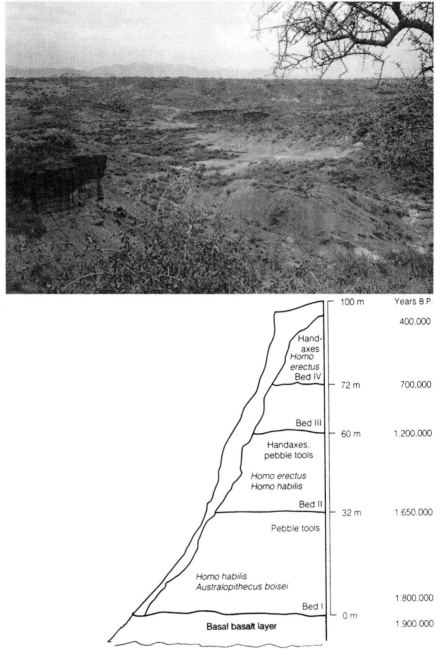

Fig. 2.7. Olduvai Gorge, Tanzania – section and panorama. Beds IV (*dark*) and III are visible in the outcrop. (Photograph by Dr. Sandra Whitney)

Reck recognized the sedimentary nature of the rocks, but noted the record of volcanism as well as indicated by the tuff deposits. The Rift Valley, an active seismic zone of extension where the African and Eurasian continental plates are pulling apart, has had volcanic activity throughout prehistory. The volcanic deposits provide lithological and geochemical signatures of individual eruptive events identified today through *tephrochronology.* The volcanic rocks also allow techniques such as radiogenic argon to be used to directly date them. Olduvai gorge's geologic sequence was delineated by Reck and then confirmed by the early use of the then new potassium–argon direct dating technique. Bed I, containing both *Australthropines* and *habilines,* was found to date to 1.95 million years (Fagan 1994). Reck's and Daly's geologic mapping of Olduvai demonstrated the utility of stratigraphy for elucidating the *provenance* or context of important archaeological facies.

In modern geology, the nomenclature of stratigraphy has seen a new terminology introduced under the rubric of "sequence" stratigraphy. Here the term formation is replaced by a new term, that of *sequence.* Sequence stratigraphy recognizes that sedimentary rock units, adjacent to basins, have generally resulted from sea level change in transgression or regression. In the former the shoreline moves landward and in the latter it moves seaward. A vertical representation of these processes shows different facies superposed according to the particular depositional environment. For example, a transgression of nearshore sands (sandstones) underlies mid-shelf facies (shales) which are overly deep sea muds (limestones) in the section. The sequence of transgression, high-stand, regression and low-stand strata, as typically seen in seismic sections, are designated as *tracts.*

A system tract is made up of *parasequences*, which are made up of sets of facies like those just described (Reading and Levell 1996). Erosional surfaces separate parasequences. A regression or abrupt basinward shift in sedimentary facies exposes a depositional sequence to subaerial weathering and creates what traditional stratigraphic nomenclature would term an unconformity. In a sequence stratigraphic description of a depositional system the "formations" are replaced by "sequences" with gaps or unconformities interpreted as flooding (ravinement) intervals of a significant temporal duration. Figure 2.8 illustrates the relationship of sea level and time in a stratigraphic sense. One can readily see how sets of deposits form in the "sawtooth" portions of the overall sea level curve.

For the archaeology of the Pleistocene sea level has played a significant role alternately exposing or flooding vast areas of continental margins. Worldwide or eustatic sea level change in the Pleistocene has been driven by glaciations where large volumes of the oceans have been sequestered or released. These dramatic shifts in sea level during regressions opened corridors and "land bridges" between continents such as Asia and North America and between Europe and the British Isles. Both humans and other biota used these connections for colonization purposes. In the lithostratigraphy of coastal areas we see this recorded in the sedimentary sequences for those periods.

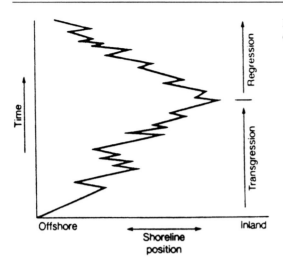

Fig. 2.8. Sequence of sea level changes in geological record.

2.12
Geological Mapping Techniques

Much of the discussion involving geomorphic mapping is appropriate to that of geologic mapping. The use of aerial and other remotely sensed data is important to the mapping of *surface* rock units. The geologic map thus shows the distribution of rock units as they appear on the surface of the earth. Like the geomorphic map the data are overlain on existing maps at a scale chosen by the investigator. The two types of maps are similar, but the geologic map uses a suite of symbols that are commonly agreed upon by geologists. Not only does the map show rock types, it also shows elevation, strike and dip of the geologic structure. Strike or line of strike is the direction or trend of a formation relative to north or south. Dip is the angle between the horizontal or the tilt of the formation. On geologic maps these appear as t-shaped symbols(†) where long lines indicate strike and short lines indicate dip. A number can be shown by the symbol to indicate the angle of dip (e.g., 25 †). As all maps *must* be oriented to the north, the angle of strike is generally deduced from the map itself.

Foliations are indicated by teeth-like symbols while multiples of this symbol indicate a reverse fault and lines with adjacent directional arrows illustrate slip-strike faults (Fig. 2.9). Chronological designations of the rock units are given by abbreviations such as "Q" for Quaternary; "T" for Tertiary, etc. Color-coding is often used to good effect in geologic maps. Contour lines allow the inference of three-dimensional arrangements of rocks and sediments.

Topographic profiles, cross-sectional illustrations or sections can be produced of outcrops of formations to show stratigraphy in the mapped area. These are not maps in the true sense, but can be used in conjunction with the surface geologic map to gain a better understanding of the rock units within the area. Structural features such as folds – synclines and anticlines, faults and

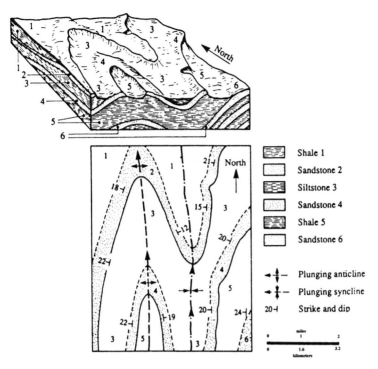

Fig. 2.9. Block diagram (*above*) and geological map (*below*) illustrating synclinal folding, anticlinal folding, strike–dip directions. (Allard and Whitney, *Environmental Geology Lab Manual*, Wm. C. Brown, Fig. 4, 1994)

unconformities can be shown in sections across the geologically mapped area. The bedding of formations – thicknesses and angularity – is also presented in this format. The profile is indicated on the geologic map as a straight line such as "AB". On the profile the horizontal maintains the scale of the surface map, but the vertical scale is more often than not exaggerated to several times that of the horizontal scale. The vertical exaggeration is a ratio of the horizontal scale divided by the vertical scale, such as a scale of a topographic map of 1:24,000 (2000 feet to the inch) divided by a scale of 200 feet to the inch yielding a ratio of 2000/20 or a 10× vertical exaggeration.

Simple rules to follow involving topographic and geologic maps include:

1. All points on any given contour have the same elevation.
2. Contours of different elevation *never* intersect each other.
3. Contour lines *never* split.
4. Contour line spacing – close versus wide – indicate steep versus gentle slopes; cliffs are indicated by closely spaced or seemingly coincident lines.
5. Contour lines close on themselves, either within or without the confines of a map.
6. A contour that closes on itself within a map indicates a *hill*.

7. A contour that closes on itself, with hachures, indicates a *depression* – the hachures point inward.
8. Contour lines curve up or "inward" – pointing "up" a valley. The lines cross streams at right angles.
9. Bold or darker contour lines reflect simple multiples of one another. Some maps indicate a multiple of 5 or 25, etc. This is termed the *contour interval* (Fig. 2.10).
10. Anticlines have their *oldest* beds or rock units in the center.
11. Synclines have their *youngest* beds in the center.
12. Anticlines incline ("plunge") toward the *closed end* of a geologic structure.
13. Synclines incline toward the *open end* of the geologic structure.
14. *Monoclinal* folds incline in only one direction. Terraces act as monoclines.
15. Contacts of rock units with a dip greater than the stream's slope or gradient "*V*"*downstream*; those with a dip less than the gradient,"*V*"*upstream*.
16. Vertical beds do not "V" or migrate with erosion.
17. Contacts migrate down-dip with erosion.
18. Fault blocks erode according to their up- or downthrown aspect – upthrown eroding faster than downthrown fault blocks.
19. Contacts between horizontal rock units are parallel to the topographic contours along those contacts.
20. The top of a map is *always* north.

2.13
Other Types of Maps

2.13.1
Paleogeographic

These maps are typically regional or broader in scale showing the distribution of ancient land and sea. In Europe and North America the most common paleogeographic maps are perhaps those showing the extent of the Pleistocene ice sheets relative to unglaciated terrain. Another commonly seen representation is past sea levels and coastlines in the Pleistocene. Unlike paleogeographic maps from more ancient epochs which rely on the plotting of rock units that are time-synchronous, paleogeographic maps of archaeological interest are based on more geologically recent mappable surficial features such as moraines and strandlines.

2.13.2
Isopach

These maps are used to illustrate the thickness of particular strata. Unlike profile illustrations, the stratigraphic unit is mapped and contoured at some

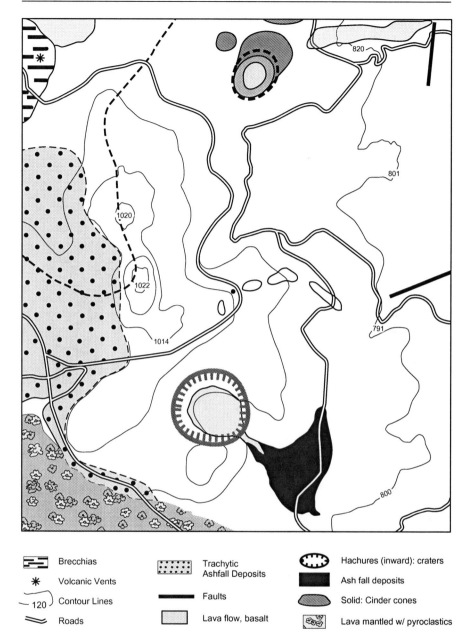

	Brecchias		Trachytic Ashfall Deposits		Hachures (inward): craters
	Volcanic Vents				Ash fall deposits
120	Contour Lines		Faults		Solid: Cinder cones
	Roads		Lava flow, basalt		Lava mantled w/ pyroclastics

Fig. 2.10. Typical geological map – contours, craters (*hachures inward*), and faults (*dark lines*). Area is northwest of Clermont-Ferrand, France – "Chaine des Puys".

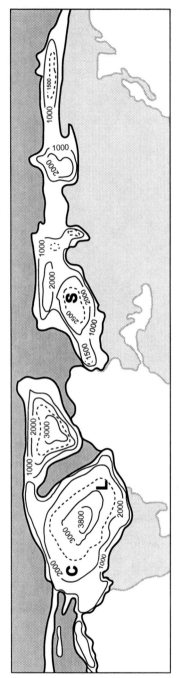

Fig. 2.11. Isopach representation of ice sheet thickness. *C* Cordillearean; *L* Laurentide; *S* Finno-Scandic. Ice sheet thickness in meters.

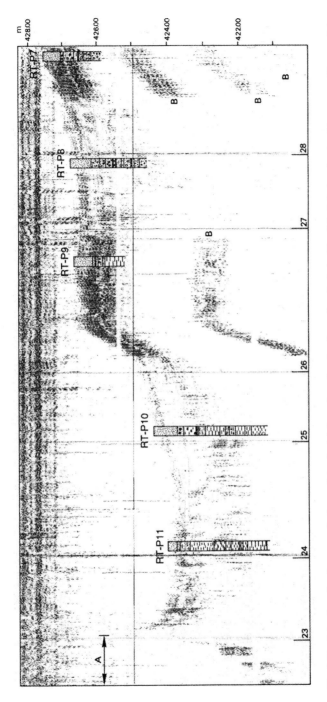

Fig. 2.12. Seismic reflection record of morainic glacial deposits taken by the author in Lake Neuchâtel, Switzerland, with core profiles overlain. (Adapted from Hadorn 1994, p. 39)

selected interval. The ice sheet thicknesses shown in Fig. 2.11 illustrate a typical isopach representation. Isopach maps generally are done for buried strata and hence, are based on core data. The use of isopach maps is infrequent in the presentation of archaeological data. As a rule when core data are available, a profile representation is done with the stratigraphic data overlain at specific sounding stations as shown in Fig. 2.12. When doing an isopach map, the upper surface is used to portray the stratigraphic unit. Isopach maps are very useful in determining the shape, direction and depth of basin-like features and buried channels.

2.13.3
Lithofacies

A lithofacies is defined as that aspect or attribute of a sedimentary unit that is a direct consequence of a particular depositional environment or process. An obvious example is a fluvially deposited unit wherein the grain shape, mineralogy, surface texture, size, etc. of the sediment reflects stream transport. Another way to visualize a lithofacies map is to think of it as an isopach map without contouring. Sometimes archaeologists will use a variant of the lithofacies map in showing the association of archaeological sites with specific landforms or soils. In the latter case, Fig. 2.13, from Kuper et al. (1977), shows the association of Bandkeramik (LBK) sites with loessic soils in northwestern Germany, west of the Rhine River valley. The lithofacies map is a qualitative representation of surficial and subsurface strata known from isolated soundings. One can assume the distribution of some strata will not be

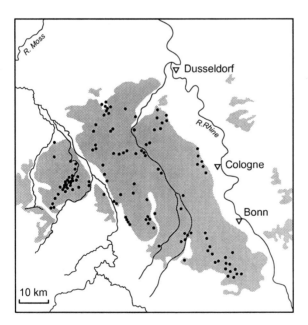

Fig. 2.13. Linear band ceramic (*LBK*) sites plotted relative to loess soils in northwestern Germany. (Adapted form Kuper et al. 1977)

uniform or continuous. This is true for most intra-site mapping and limits the applicability of the technique. Where soundings are of sufficient density the use of lithofacies maps has merit. Where a clear association of archaeological material with a stratigraphic unit is demonstrated, then the lithofacies mapping of the unit based on pedological/sedimentological features assists in bounding the site for a specific occupational period. This method works very well in shallow (<1 m) depth sites that can be quickly sampled with hand-coring equipment (see Chap. 4, this Vol.).

3 Geophysical Techniques for Archaeology

3.1
Introduction

The objective of this chapter is not to make the reader adept in the theory and application of geophysical techniques used in archaeology today. The goal is to present a relatively comprehensive survey of "shallow" geophysics, ranging from electrical – resistivity and conductivity – to magnetic and radar methods. The methods to be discussed are the ones most commonly used in archaeological geophysics. Less frequently used methods – seismic, gravity, thermographic, induced polarization, self-potential – to name some of the other methods used in shallow geophysics, will not be examined in any detail. Many of these other methods have achieved interesting results, in regard to archaeological prospection, but for a variety of reasons they remain of marginal interest to archaeology today. As with all the geophysical methods utilized by archaeology, *none* were developed with archaeological prospection in mind. As is true with most, if not all, of the various methods used by archaeology, these methods have been borrowed and adapted to fit archaeological goals.

In this survey we shall examine basic principles of the specific geophysical technique under discussion, then proceed to a brief discussion of the technique's importance to archaeology. Finally, a hypothetical or actual case example will be presented to help the readers decide for themselves of the utility of the method to their needs. Mathematical formulae and equations are part and parcel of geophysics, but in this chapter we shall only examine those most basic and germane to the method. There are a host of works, some referenced in the following paragraph, such as Irwin Scollar (1990), which do provide in-depth summaries of the physics and formulae involved in the use of these geophysical methods. Excellent basic textbooks on geophysics, such as those by Kearey and Brooks (1991), Burger (1992) and Sharma (1997) are readily available to those readers who seek a more in-depth understanding of the principles involved.

The textbooks (Clark 1990; Scollar 1990), major reviews (Weymouth and Huggins 1985; Clark 1986; Weymouth 1986; Wynn 1990; Weymouth 1996), and dedicated journals – *Archaeological Prospection* (1994) published in the last decades of the twentieth century are indicators of the increased importance of geophysical techniques for archaeology. Excellent historical background reviews appear in some of these works as well as elsewhere. Without digressing into historical perspective, it is important to remember that the use of geophysics in archaeology began in earnest after World War II, first in Europe, then elsewhere. The sequence of development proceeded from the use of soil resistivity methods by British workers (1950s) to magnetics by the Germans and Italians (1960s) and the development of ground-penetrating radar (GPR) by Americans (1980–1990s). Other geophysical techniques have found their way into archaeological use, but to a lesser degree than those just mentioned. These latter methods include induced polarization (IP), conductivity (EM), magnetic susceptibility, microgravity, thermography, seismic reflection/refraction and electrical tomographic imaging.

Since the 1960s, the International Symposium on Archaeometry has prominently featured archaeological prospection in its conference programs. The increasing inclusion of geophysics into archaeology has seen the use of interesting terms to describe this. Wynn (1987, 1990) proposed the term "archaeogeophysics" for all ground and airborne geophysical methods applied to archaeology. In the strictest sense, most geophysical methods used today are terrestrial ("ground") in nature, although the use of airborne or space-based sensing technology has produced significant archaeological finds ranging from buried, "desert" rivers (Wendorf et al. 1987; Wiseman 1996) to Mayan field systems in tropical forests.

There is an additional use of geophysical methods that Wynn does not specifically include in his 1990 definition although he discusses it in his review article at some length – seismic reflection in marine archaeology. Today, seismic profiling – analog and increasingly digital in nature – coupled with magnetometers and side-looking/scanning sonars are standard marine search procedures conducted on archaeological sites ranging from shipwrecks to drowned terrestrial landforms of archaeological interest.

The suite of geophysical instrumentation developed or adapted for archaeological prospection is indeed impressive. With the concurrent usage of geophysics in mineral prospection, environmental studies and geotechnical engineering, the development of instruments that are improved and increasingly affordable has enhanced their availability. In the past, even as late as the past decade, a common caveat used in explaining the lack of the use of archaeogeophysical techniques was economics. Cost has been a factor, but perhaps a lack of an informed cadre of archaeologists was more at fault for the misconception than the former. Many, if not all, major archaeological projects allocate financial resources according to methodological goals. Before the radiocarbon dating and subsequent chronometric techniques became commonplace and their utility understood by the archaeological community, few project resources were earmarked for their use. Today, it is almost unheard of to not include some kind and amount of chronometric determination in an archaeological project. Beyond venues like that of the International Archaeometry Symposia, research and application of archaeological prospection has received more time at less technically focused professional meetings. Professional conferences in the earth sciences, archaeology and remote sensing commonly include papers on the theory and use of geophysical methods in archaeology. Add to this favorable trend the use of workshops in the "hands on application of these techniques " it is more and more commonplace to see them deployed in a full spectrum of archaeology ranging from classical Mediterranean, to African, Asian and American sites.

Academic training in archaeological prospection has evolved as well. Important examples can be found in European (Oxford, Bournemouth, Zürich, CNRS-France, Warsaw, Vienna and Rome) academic settings together with the North American universities: UNAM (Mexico City), University of Georgia, University of Pennsylvania, Florida State University, University of Arkansas and Louisiana State University. As today's students become skilled in theory and practice, the application of geophysics in archaeology is assured.

3.2
Electrical Methods – Resistivity

Resistivity methods have "pride-of-place" in archaeological geophysics by virtue of being the first geophysical method used by archaeologists (Aitken 1974). These early efforts, after World War II, consisted of much trial-and-error, with multiple, galvanic electrodes used as a rule. The first applications were what we will call "profiling" studies in that the electrodes – simple steel rods in most cases – are advanced over the site at a fixed interval between the probes – Wenner or Schlumberger methods (Aitken 1974). A rough rule-of-thumb, still followed today, holds that the electrode separation is equivalent to the profiling depth or that a 1 m electrode separation will detect buried features at 1 m depth in the site.

Electrical resistivity or resistance survey methods measure electrical flow through soils or sediments. Most earth materials conduct currents, poorly at best, but vary nonetheless, according to mineralogic makeup and water content. Rocks are poor conductors. Some of the metallic or igneous varieties, such as the sulfide minerals, basalt, gabbro, diorite, and magnetite are better than the sedimentary or metamorphic rocks.

Table 3.1 lists resistivity values of common sediments, rocks and minerals. Of the sediments, clays are the best conductors and clay-rich soils have lower resistance to electrical flow than say, quartzitic or carbonate types. The difference in the resistance or resistivity, as it is most commonly called, as measured by galvanometers or ohm/voltmeters, is used to detect and characterize buried features of an archaeological nature. The fundamental physical relation involved in resistivity is Ohm's law:

$$R = V/I$$

Table 3.1. Resistivity of common materials

Material	Frequency at measurement (cps if not DC)	Resistivity (ohm-cm)
Minerals		
Galena	–	0.5–5.0
Pyrite	–	0.1
Magnetite	–	0.6–1.0
Graphite	–	0.03
Rock salt (impure)	–	3×10^3–5×10^5
Serpentine	–	2×10^4
Siderite	–	7×10^3
Igneous rocks		
Granite	–	10^8
Granite	16	5×10^5
Diorite		10^6
Gabbro	–	10^7–1.4×10^9
Diabase	–	3.1×10^5
Metamorphic rocks		
Garnet gneiss		2×10^7
Mica schist	16	1.3×10^5
Biotite gneiss	–	10^8–6×10^8
Slate	–	6.4×10^4–6.5×10^6
Sedimentary rocks		
Chattanooga shale	50	2×10^3–1.4×10^5
Michigan shale	60	2×10^5
Calument and hecla conglomerates	60	2×10^5–1.3×10^6
Muschelkalk sandstone	16	7×10^3
Ferruginous sandstone	–	7×10^5
Muschelkalk limestone	16	1.8×10^4
Marl		7×10^3
Glacial till	–	5×10^4
Oil sand	–	4×10^2–2.2×10^4

Rock type	Resistivity range ($\cap$-m)
Consolidated shales	20–2×10^3
Argillites	10–8×10^2
Conglomerates	2×10^3–10^4
Sandstones	1–6.4×10^8
Limestones	50–10^7
Dolomite	3.5×10^2–5×10^3
Unconsolidated wet clay	20
Marls	3–70
Clays	1–100
Alluvium and sands	10–800
Oil sands	4–800

In Ohm's law, R is the resistance (measured in ohm, Ω), V is the voltage (measured in volt' V) and I the current (measured in ampere, A). One ampere is equal to one coulomb of (electron) charge moving in 1 s through a potential of 1 V. For a current to flow there must be a difference in the voltage potential, also termed bias. Resistivity and resistance are not totally synonymous terms. Resistivity is specific resistance, expressed as ρ in the unit of ohm-meter, Ωm. It is the resistance to current flow over some distance and is proportional to the area. In a resistivity survey, we measure the apparent resistivity, ρ_a, which is the "true" resistance across a nonhomogeneous media like the earth. The apparent resistivity is expressed with a geometric correction factor that is particular to a particular electrode spacing or array. The geometric factor for the Wenner array, the most commonly used array, is $2\pi a$, where a is the electrode spacing and 2π considers the half-space volume over which the current flows. With this in mind, we rewrite Ohm's law, for the Wenner array as:

$$\rho_a = 2\pi a\, V/I$$

A material with significant resistance to current flow is an insulator while one that allows current flow is a conductor. Glass and silicate minerals are insulators while the metals are conductors.

The conductance of current through soil is greatly influenced by water content. Water alone is not necessarily as good a conductor as distilled water is a good insulator, whereas tap water is a conductor. Why? Distilled water is free of any mineralogic salts and other ions where tap water has charge carrier ions such as sodium, chlorine, etc. that *conduct* electrical current. This property is termed *ionic conductivity*. In sediments, the number of ionic elements is much greater than simple water. It is the water that mobilizes the ions to form the electrolytic soil/water solution. When an electrical potential is applied across a soil volume with electrodes, the material and moisture will act first as charge carriers. After a period of time the soil will impede the motion of ions so that they accumulate at grain boundaries wherein the ground acts like a battery. This is particularly true where direct current (DC) is applied to the electrodes. To avoid unwanted polarization effects, resistivity devices use alternating current (AC). An alternating current source, regulated by an oscillator, varies the sign of the current in +1/2 to –1/2 cycles such that the ground "battery" cannot build up charge without subsequent reversal and discharge. As a result there can be no net polarization. The early two-electrode arrays were prone to another problem known as "contact resistance" which comprises a large part of the total resistance. Workers overcame this problem by adding two additional electrodes, equally spaced apart and the four-electrode array was born.

As stated, resistivity is the most venerable of geophysical techniques used for archaeology (Aitken 1974), and will continue to be used into this millennium as a productive prospection method. Over a half-century of experience in both archaeology and geology, resistivity has provided a solid conceptual as well as applied basis. To remain a productive geophysical methodology for archaeology, advances in electrode design such as nonpolarizing electrodes; together with current reversal circuitry, low power AC batteries, and low induction cables have all contributed to a more efficient modern resistivity technology. These hardware innovations have been coupled with the advances in computer design to make modern instruments more compact and flexible. Data are in digital formats making storage manipulation and modeling by geophysical software all the more routine. The most often heard problem – lack of speed in field survey – has been addressed, at least in archaeology, by the modern double-dipole array instruments and the multi-electrode array instruments. The multi-electrode (25+) cable arrays used with computer-driven control units allow the rapid collection of field data in a host of array configurations – Schlumberger, Wenner, double-dipole, etc.

3.3
Resistivity Arrays

In archaeological prospection four-electrode arrays are most commonly used. We will now review them in order of popularity. The Wenner array is the "standard" against which other arrays are assessed (Milson 1997).

3.3.1
Wenner Array

The high impedance measurement circuit used for Wenner potential electrodes (P_1, P_2) reduces contact resistance by orders of magnitude because it draws little current. It has been found that electrode size, depth and geometry are of little importance, but electrode spacing, as we have said, is more so. With the Wenner array this relationship is fairly direct with the depth of profiling roughly equivalent to electrode spacing. Depth and spacing are generally treated as equivalent.

The diagram in Fig. 3.1a illustrates the spacing, a, is equal and set to the depth of interest for the specific traverse. To traverse a site, the array is advanced by increments equal to "a" with readings made. The direction of the survey will influence the appearance of most anomalies. While it is time-consuming it is sometimes useful to re-survey at right angles to the direction of the original survey. For sounding surveys the electrode spacing is expanded by increasing increments of "a" = 1, 2, 4, 8 . . . n. It is the easiest of the four-electrode arrays to conceptualize. One simply lines up the four electrodes at an equal interval "a", drives a current, reads the resistance and then moves the electrodes, in sequence, by a predetermined interval "a". This is repeated until the area of interest has been completely traversed on an X–Y grid.

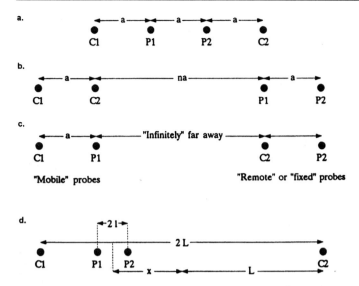

Fig. 3.1a–d. Resistivity arrays. a Wenner array; b double-dipole array; c twin electrode array; d Schlumberger array – exists in several variants

3.3.1.1
Archaeological Application of the Wenner Array: Resistivity Profiling

All of the arrays discussed in this section can be used for archaeological pro-filing studies and indeed they have. The profiling method simply requires the traverse of an archaeological site using a pre-selected offset and sampling interval. This is termed the "constant-spread" traverse.

Example. Sixteenth century Spanish mission – St. Catherine's Island, Georgia.
Lost since the late seventeenth century, the Franciscan mission of Santa Catalina de Guale was re-discovered in 1981 by a team using magnetic survey (Garrison et al. 1985; Fig. 3.2). Subsequent geophysical study of this large Spanish-American Indian mission town utilized resistivity and ground pen-etrating radar (Thomas 1998a). In the resistivity study, led by Mark Williams (1984), a Wenner configuration survey, using a probe separation of 1 m, along an east–west traverse of 20 m, resulted in 21 resistance readings. Figure 3.3 was produced from this data around an area of the mission kitchen or *cocina* structure, previously excavated. The resistivity plot helped in further deter-mining the limits of this structure (Thomas 1987).

Resistivity Sounding. This category of resistivity measurement is typically described as a sounding method (Burger 1992), but that is only if the center of the array is kept at a fixed location and the electrode spacing (a) in the Wenner method, is sequentially increased. This common variation is to

Fig. 3.2. Aerial photograph of excavated area of Mission Santa Catalina de Guale as it was in 1984. The top of the photograph is north, with the *white, fiduciary crosses* spaced on 20-m grid intervals. The linear N–S features and block areas are the excavated portions of the site. The mission church is *lower-center* while the mission/presidio well is *upper right* and the kitchen is *upper-center*. (Photograph by Dennis O'Brien reproduced with the kind permission of the American Museum of Natural History)

simply repeat the constant spread traverse at increased increments of (a). Rather than construct a resistivity *pseudosection,* described in the following sections, each (a) spacing traverse's data, for that level, can be viewed independently of preceding or subsequent traverses. This data can be modeled in an areal contour plot rather than in a depth section. Multi-level plots can be made treating each level as a "resistivity surface" (Herbich 1993). We shall provide an example of this type of profiling in the discussion of the Schlumberger array.

3.3.2
Double-Dipole Array

In principle, the double-dipole array mimics induced polarization (IP) techniques in array configuration and measurement. The principal difference in

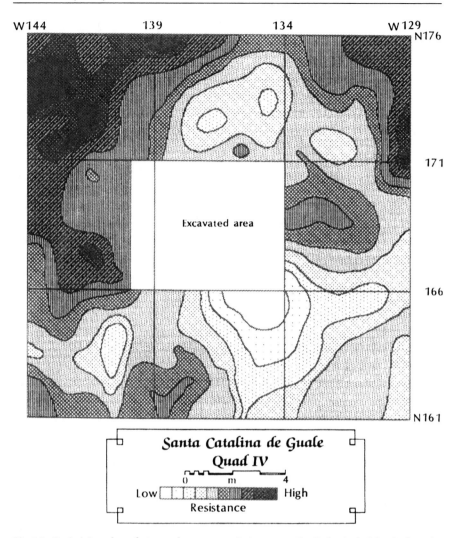

Fig. 3.3. Resistivity plot of sixteenth century mission area. St. Catherine's Island, Georgia. (Courtesy of the American Museum of Natural History)

the two techniques is the method of applying and measuring potentials – the double-dipole array applies a continuous potential while IP relies on induced, remnant potentials transient in the soils. For a discussion of IP methods see Wynn (1988). The double-dipole array has the potential electrodes separated at some distance from the current electrodes (C_1, C_2) while the electrode spacing (a) being the same for both sets of electrodes (Fig. 3.1b) involves a geometric factor that differs from Wenner and Schlumberger arrays. Here, the parameter n is the distance between the current electrodes.

$$\rho_a = \pi a \, n(n+1)(n+2) V/I$$

Williams (pers. comm.) touts the double-dipole array over that of the Wenner array in archaeological prospection, primarily because of the location of the measured maximal value of apparent resistivity. With the Wenner array, the profile of a resistive feature takes on a double-peak shape and is reminiscent of the magnetic dipole shapes discussed later in this chapter with the maximal value for ρ_a, over the feature, reduced because of electrode spacing and type.

This "double-peak" signature is not seen in plots of double-dipole data as the most sensitive location for measuring apparent resistivity is measuring at ηa (Fig. 3.1b). Two disadvantages of the double-dipole array are: (1) overall sensitivity and (2) comparability of reading to those of other arrays. This is true for either profiles or soundings. The double-dipole array has become popular in both archaeological and mineralogical prospection using sounding data to produce 2-D inversions and pseudo-sections (Griffiths and Barker 1994; Milson 1996). Both Wenner and Schlumberger arrays produce more reliable, deep soundings.

3.3.2.1
Archaeological Application of the Double-Dipole Array

Example. Williams and Shapiro (1982) carried out a double-dipole array survey at the colonial French-British fort of Michlimackinac. The eighteenth century site has seen extensive archaeological research, particularly the fort location (Stone 1974). Somewhat less-well known was the location of the village associated with the fort. Williams and Shapiro did their study in an area of roughly 0.5 ha, using an electrode spacing of 1 m. This methodology resulted in a data set of over 5000 resistance readings. Figure 3.4 illustrates the results in a contour plot. The plot is typical of shallow, single-component archaeological sites such as a colonial village occupied only a brief time (1760–1781). Rather than producing linear features such as seen in settlements with well-defined streets and buildings which, in turn, produce buried remains with significant contrasts in their resistance, short-term settlements typically do not. As Williams and Shapiro demonstrate in these latter sites, resistive anomalies and features are not greatly contrastive and are generally scattered across the site. Patterns are random-like and it is hard to tease out lineations or geometric shapes that represent structures. Add to this picture, the fact that the village site had been a public park for decades. Modern debris, trenching for conduits and pipes, and other twentieth century disturbances, have only added to the elements that will produce spurious resistive readings. Nonetheless, unlike at the sixteenth century Spanish mission, occupied for a century, the resistivity data from Michlimackinac indicate subsurface features that, in large part, should be of an archaeological nature.

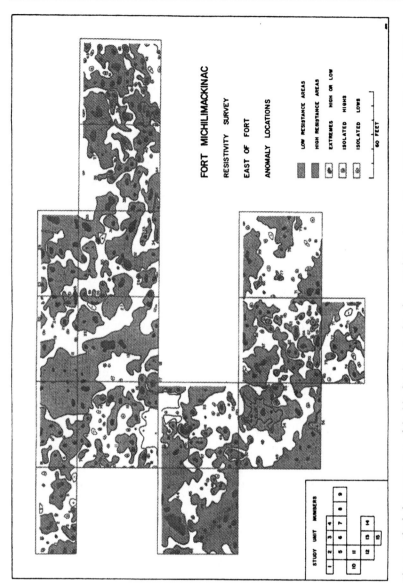

Fig. 3.4. Shaded contour map of double-dipole array data from Michilimackinac village. (Williams and Shapiro 1982)

3.3.3
Two/Twin Electrode Array

Milson (1996) characterizes the modern variant of this array (double-dipole) as the one most popular in archaeological prospection today. The successful modern variant is a DC system with a fixed inter-electrode distance (a) that allows the unit to be operated by a single person without cables to caddy (Fig. 3.1c). The use of DC with its polarizing effect has been ameliorated by use of a switching circuit to minimizing charging.

The twin electrode array was especially designed for archaeological prospection (El-Gamili et al. 1999). As Fig. 3.1c indicates, it is a Wenner array divided into two parts with a very large separation between the current and potential electrode pairs (El-Gamili et al. 1999). Aspinall and Lyman (1970) have shown that this arrangement allows for the same penetration of a Wenner array with twice the separation between the electrodes. The thing that commends the twin electrode array to archaeological prospection is its compact nature. The so-called "mobile probes" are on a fixed frame with a cable connection to the other electrode pair located at a fixed point some distance away (greater than 30 times the electrode spacing as a rule). Another benefit of the large separation in the array halves is a reduction in background resistance. Current practice in Great Britain and elsewhere in Europe favors the use of the twin electrode array. It is quick – up to 6000 readings per day – and coverage is comparable to that of a magnetometer survey – a hectare per day.

3.3.4
Schlumberger Array

Conceptually, the Schlumberger array is as well understood as the Wenner array and is perhaps more used than that array for sounding work. On the first inspection, the two arrays appear similar with the potential electrodes in the array's center and the current electrodes on the ends. The potential electrodes are fixed in their spacing such that a minimum factor of 2.5× exists between their inter-electrode distance, 2l, compared to that of the current electrodes, L (Fig. 3.1d). The array can be moved as an ensemble for a profile survey or the current electrodes moved separately increasing L, by logarithmic steps, to produce sounding data. The inner electrodes are not moved until voltage levels are not measurable. The apparent resistivity for this array is calculated as follows:

$$\rho_a = \pi[(L^2 - l^2)V]/2l$$

Williams (1982:113; 1984:14) considers a 1-m-depth/electrode spacing adequate for most archaeological resistivity work. This being the case the use of a rigid frame resistivity meter mounting fixed-distance electrodes (a), and a data-logger or portable computer is in keeping with most profiling needs of an archaeological nature. Calculation of the apparent resistivity for the two-electrode array is the same as that for the Wenner array.

3.3.4.1
Archaeological Application of the Schlumberger Array

Example. The Prehistoric flint mines of the Holy Cross Mountains, Poland. Herbich (1993) conducted extensive resistivity surveys at the prehistoric flint mines at Wierzbica, Zele, Krzemionki and Polany. Using the Schlumberger array, Herbich and his coworkers carried out multilevel profiling studies of the respective mine sites in an attempt to determine the depth, density and nature of the mine works. As each survey block was completed the array was expanded to provide deeper and deeper penetration of the same survey block. After the data were processed, its display in multilevel plots (Fig. 3.5) is informative of the mining activity at different depths as well as its location. One can easily see the separation in the zones of shallow and deep mining (Herbich 1993).

3.4
Vertical Sounding Methods in Archaeology

Electrical sounding has been called "electrical drilling". As such it can be thought of as a type of coring device to get at the depth as well as the distribution of archaeological features, i.e., pits, hearths, remains of structures. As we have seen in the preceding example from Herbich's study of ancient mining, the Schlumberger technique provides excellent profile data. It can also be used for vertical soundings, involving the expansion of the spread of current probes AB around a central point with constant position of potential probes MN, giving information on the number of horizontal resistivity interfaces at that point (Fig. 3.6). The result reflects the properties of a vertical section or *pseudosection* of the subsurface, but:

- The quality of the interpretation decreases radically with depth, since the properties of larger volumes of soil are being sampled by each expanding reading;
- Each datum point will include significant contributions from volumes of soil not directly under the array – the pseudosection compresses all such effects onto a two-dimensional display of apparent resistivities onto an arbitrary vertical grid;
- The pseudosection, whether displayed as a block of numbers, a gray-scale image, or in contoured form, is not a direct representation of the subsurface and any interpretation as an image can be extremely misleading – the depth scale is completely arbitrary. In the past processing, filtering or smoothing of the data could not seem to overcome this fact (Szymanski

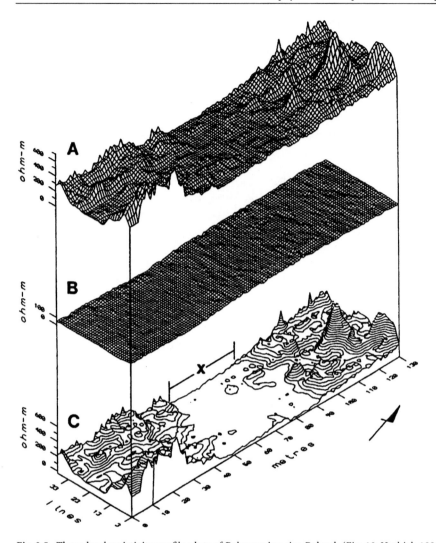

Fig. 3.5. Three-level resistivity profile plots of Polany mine site, Poland. (Fig. 10, Herbich 1995)

and Tsourlos 1993), but the use of inversion techniques have improved results significantly (Sasaki 1989; Barker 1992; Tripp et al. 1984).

Today, with double-dipole data, inversion of the pseudosection data gives a 2-D tomographic image. It is now possible to get accurate depth and location using the inversion process (Sharma 1997). By starting with an initial model – the pseudosection – which is essentially a "guess" as to what the subsurface looks like, the inversion then changes or "iterates" the model data until the model agrees with the collected data within accepted error. These are called "error-minimization" routines. It is analogous to the technology used in

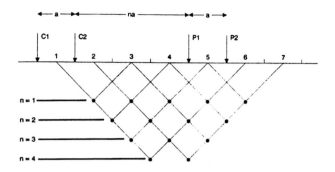

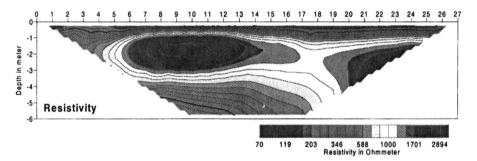

Fig. 3.6. Typical resistivity sounding array (*above*) and a representative pseudosection (*below*). The pseudosection maps karst limestone near Austin, Texas. (Courtesy AGI, Inc.)

medical MRI and CAT scans. Barker (1992) developed an algorithm based on the finite difference method that inverts the pseudosection and converts the apparent resistivity measurements to "true" resistivity – an immediate gain is the adjustment of depth variation. For a detailed discussion of the inversion method see Szymaski and Tsourlos (1993), Noel and Xu (1991) as well as Barker (1992).

Having given the reader these caveats, many current articles describe the use of electrical sounding and in particular suggest the use of pseudosections to aid in the interpretation of the subsurface archaeology (Dalan et al. 1992; Szymanski and Tsourlos 1993; Aspinall and Crummet 1997). Griffiths and Barker (1994) have used a computer-controlled electrode array along a fixed-length cable so that the sequence, spacing and array configuration are selected and manipulated by the software resident in the control unit (Fig. 3.7).

3.5
Electrical Methods – Electromagnetic/Conductivity

The reciprocal of resistivity is conductivity and is written as the Greek sigma σ, so:

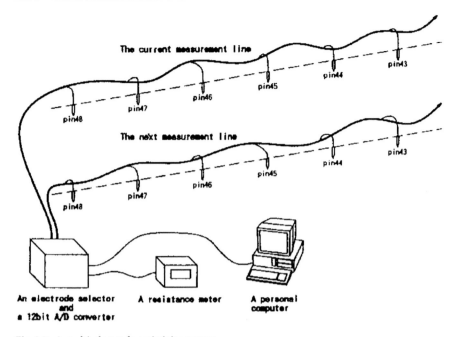

The current measurement line

pin43
pin44
pin45
pin46
pin47
pin48

The next measurement line

pin43
pin44
pin45
pin46
pin47
pin48

An electrode selector
and
a 12bit A/D converter

A resistance meter

A personal
computer

Fig. 3.7. A multi-electrode resistivity system

$$\sigma = 1/\rho$$

Electromagnetic (EM) induction, avoided in resistivity measuring systems, is utilized to measure the conductance in the subsurface. Like resistivity differences in conductivity measured in the subsurface may have as their source objects or features of archaeological interest. Induction of eddy currents is a consequence of a primary dipole field whose magnetic field lines flow through a conductor producing the weak secondary currents (electrical). Conductive minerals like clays and metallic oxides in soils enhance induction by what is termed mutual inductance (Milson 1996).

A Slingram-type conductivity meter uses a source current flowing through a wire coil loop producing an alternating low frequency magnetic field (i.e., between 200 Hz and 20 kHz), while a second coil, at a distance L, measures two different components of the magnetic field – the in-phase and quadrature components. The in-phase component is a measurement of the magnetic susceptibility of the soil and that of quadrature a measurement of the soil conductivity. In-phase measurements are like those of a "metal detector" only much more sensitive. Quadrature measures are those of electrical properties of the soil.

The response of the conductivity meter is complex: the secondary field is not in phase with the primary field and this increases with the conductivity of the ground. The presence of this secondary field is a direct consequence of the magnetization induced by the primary field. High frequency waves attenuate first and quite rapidly with depth. As a consequence, frequencies above 1000 Hz are rarely used in devices for archaeological prospection. A semi-empirical formula relates soil resistivity to frequency as follows:

$$h\sqrt{f/\rho} = 10$$

where h is the depth, f the frequency and ρ the resistivity. For instance, if we solve for f with $\rho = 10^4$ ohm-m and h = 10 m, then f should be 100 Hz.

With the continuous wave (CW) frequency domain EM devices such as the Slingram, Geonics EM-31 and EM-38, the primary transmitter coil puts out a sinusoidally varying current at a set frequency which serves as a reference for the secondary (receiver) coil whose current characteristics are compared to the primary. The amplitude and phase angle components of the current are compared, e.g., the in-phase and quadrature (out-of-phase) components. These are read directly on a digital meter on the device. The principal advantage of the conductivity meter is *speed*. These units can cover the area of a site as quickly as the magnetometer. As impressive as the new twin electrode resistivity units are, they are no match for the ease of survey that the conductivity and magnetometer provide to archaeological prospection. It should be noted that the noncontact nature of the conductivity measurement usually results in a reduced sensitivity, ordinarily by a factor of 5, when compared to resistivity measurements, but this is usually not important in archaeological cases.

Newer, Multifrequency Devices. These instruments, such as the GEM-300 manufactured by GGSI, Inc., regularly operate over a wide range of frequencies – 100 to 20,000 Hz. As we recall the effect of frequency on the soil, generally speaking, the higher frequencies "see" all soil as conductive and, therefore, the depth of survey is quite shallow. Newer electronic circuitry and coil geometries allow the use of higher frequencies which are particularly good at isolating shallow conductive features such as those interesting to archaeology (Fig. 3.8). These newer multi-frequency devices allow for the acquisition of large, digital data sets that can be plotted and compared as to the presence of conductive features and their "depths". True depth estimations are problematic in EM survey results. With units like either the EM-31 or EM-38 the rough relationship between coil separation (1.2× coil separation) and depth can be used. The E-31 has a maximum depth of survey of about 6 m while that of the EM-38 is only 1.5 m. This problem will crop up again in our discussion of another EM-based device: the ground-penetrating radar (GPR).

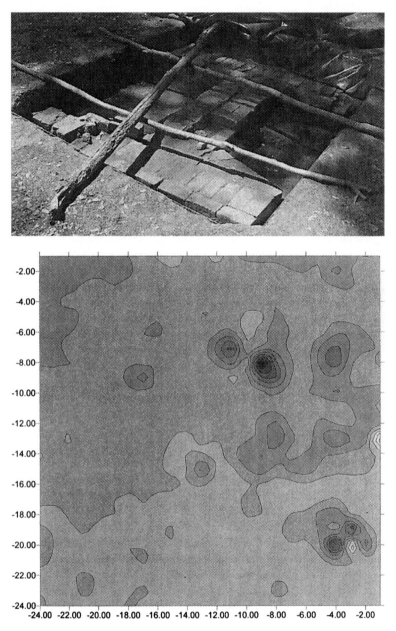

Fig. 3.8. Excavated foundations of chimney (*above*) and plot of soil conductivity (*below*), Scull Shoals village, Georgia, USA. (Photograph by author)

3.6
Application of Electromagnetic Methods

Electromagnetic surveys are done as traverses. Like resistivity data, the results are typically presented as profiles or surface plots, either plan-view or isometric.

Example. Cahokia Grand Plaza, near Monk's Mound, East St. Louis, Illinois, displayed in Fig. 3.9, is one of the most important archaeological sites in North America.

Holley et al. in their 1993 study of the Grand Plaza have used EM survey in the central quarter or precinct of this prehistoric urban center. Rising to a cultural climax between the twelfth and fourteenth centuries, this Pre-Columbian center has given the Mississippian Culture its imprimatur. The complexity of Cahokia is daunting to the many archaeologists who work there. The increased usage of geophysics can help to unravel Cahokia's questions of scale and complexity.

3.7
Magnetic Methods

The degree to which a material becomes magnetized in an applied magnetic field H is called its magnetic susceptibility. Later in this chapter, we will discuss the use of this property of rocks minerals and sediments for archaeological purposes, but for the moment we will use it to examine how magnetism is used in archaeological geophysics. How a material is magnetized is termed magnetic induction. Magnetic susceptibility, χ, is defined as:

$$\chi = M/H$$

where M is the induced magnetism and H is the total magnetic intensity of the applied field.

χ is a dimensionless quantity in the centimeter-gram-second (cgs) system and generally measured for a volume or mass. The above relation describes volume susceptibility and when divided by the material's density, χ/d, the mass susceptibility is obtained. Magnetic materials vary in susceptibility from weakly negative to strongly positive. Magnetic prospection depends on an observed contrast in the susceptibility of materials. Almost all magnetic features in archaeological sites are differentiated by the amount of the magnetic mineral magnetite, Fe_3O_4 present. Magnetism is characterized as either diamagnetic, paramagnetic so ferri(o)magnetic. Briefly, these various terms describe the magnetic response at the atomic level.

Diamagnetism arises from the distortion of an atom's electron orbits in a magnetic field. Always present, this form of magnetism is very weak (about

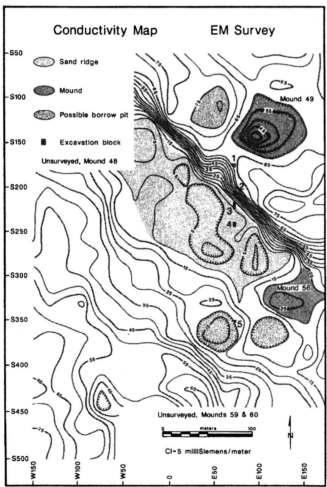

Fig. 3.9. Cahokia Grand Plaza (*above*) and conductivity plot (*below*). (Plot from Holley et al. 1993, Fig. 3)

1 nT). When an atom has unfilled electron shells, valence or otherwise, they will demonstrate paramagnetism individually. This latter point is important. By behaving individually, in bulk arrangements (crystals, glass, liquids), there is no net magnetic behavior or moment. Paramagnetic materials demonstrate weak magnetism in applied fields.

When a material's atoms with unfilled outer shells behave collectively or coupled, the magnetism can be ferromagnetic or anti ferrimagnetic, depending on orientation of magnetic moments. Iron (Fe), cobalt (Co), nickel (Ni), their alloys or minerals are powerfully attracted to magnetic fields ("magnetic induction"). As a result, ferromagnetic and ferrimagnetic materials are the most prevalent in producing strongly magnetic features in archaeological sites. Therefore, we characterize materials on their ability to "acquire" magnetism or their magnetic susceptibility. Typical susceptibilities are shown in Table 3.2. The reader will note that χ in this table is written in Gaussian or cgs units. Volume magnetic susceptibility is generally regarded as dimensionless while the magnetic susceptibility per mass is a more useful quantity for use in magnetochemistry of archaeological materials and is given the units of reciprocal density. Rocks and artifacts such as ceramics can acquire a permanent net magnetism termed remnant magnetism. There are various ways this can occur, such as heating, chemical activity, and pressure. The remanent magnetism so acquired, by whichever physical means, is relatively permanent over geological time and forms the basis for paleo- and archaeomagnetic dating methods. Interaction of the induced and remanent magnetic components yield the net magnetism of a material we observe in the earth's field. To measure this net magnetism magnetometers are utilized.

Table 3.2. Magnetic susceptibilities of common rocks and minerals (cgs units)

Top soil	10^{-4}
Clay (unfired)	10^{-5}
Altered ultrabasic rocks	$10^{-4}-10^{-2}$
Basalt	$10^{-4}-10^{-3}$
Gabbro	10^{-4}
Granite	$10^{-5}-10^{-3}$
Andesite	10^{-4}
Dolerite	$10^{-2}-0.1$
Rhyolite	$10^{-5}-10^{-4}$
Shale	$10^{-5}-10^{-4}$
Schist and other metamorphic rocks	$10^{-4}-10^{-6}$
Most sedimentary rocks	$10^{-6}-10^{-5}$
Limestone and chert	10^{-6}
Pyrrhotite	$10^{-3}-1$
Magnetite	$10^{-3}-2.0$
Chromite	$10^{-4}-1$
Haematite	$10^{-6}-10^{-5}$
Pyrite	$10^{-4}-10^{-3}$

3.8
Magnetometers

Magnetometers, unlike conductivity and resistivity devices which actively generate electric fields are "passive" instruments. By comparison, the simplest device for measuring magnetic fields is the compass. While this hardly seems to qualify this important directional tool as a prospection instrument, it was used to successfully relocate the sunken Civil War ironclad, *USS Cairo* (Bearrs 1966). Geophysicists have developed several magnetometers for use in prospection. These include the fluxgate magnetometer, proton precession magnetometer and the optically-pumped magnetometer.

Magnetometers measure the sum of the induced and remnant magnetization in some absolute or relative ratio to the earth's field depending on the instrument. Fluxgate models measure this along a single axis while the other types measure the total or absolute field – the sum of all axes of the magnetic field at a location. The single axis fluxgate readings can be subtracted from the total field to obtain the remnant component that is often directly representative of archaeological features. Absolute or total field devices, proton precession or optically pumped systems, are also used to detect local magnetic materials, artifacts, etc.

3.8.1
Fluxgate Magnetometer

This magnetometer is simply a magnetic metal around which two coils, primary and secondary, are wound to form a transformer. In operation the introduction of AC current in the primary coil produces (induces) a magnetic field in the core metal as a secondary voltage or waveform. Any external field, along the axis of the core can alter the waveform proportional to the strength of the magnetic field at a sensitivity of 1 part in 50,000 nT, i.e., 1 nanotesla (nT). If the field vector is off the axis of the fluxgate device then it is not as readily detected. Fluxgate devices are popular in Great Britain and the continent (Milson 1996; Marshall 1999; Neubauer 1999).

A three-component fluxgate magnetometer has been described by Kamei et al. (1992). Coupling two of these devices creates a gradiometer wherein both the total and gradient fields can be measured. The magnetometer can measure all three components of the total magnetic field's, X, Y, and Z orientation. Because of the complexity of the sensor(s) output a microprocessor is needed. In practice the X-axis is fixed to north and the Y and Z fluxgate sensors held in some fixed orientation to X. The data can be displayed as: (1) the north–south component (X), (2) the east–west component (Y) and (3) the vertical component (Z).

3.8.2
Proton Precession Magnetometer

This type of magnetometer is perhaps the most common field system used in archaeological prospection. Conceptually, it is somewhat more complex than the fluxgate design. As the name implies, the measurement of the magnetic field strength is through use of proton (the hydrogen nucleus) precession. Precession is a property of electrons and protons due to its spin (Leute 1987). An individual proton will align, as its spin generates a very weak magnetic dipolar moment, along the direction of a magnetic field such as the geomagnetic field of the earth.

Protons as in a fluid – water, gasoline or alcohol – can act in a collective manner similarly to that of ferromagnetic domains. A proton precession magnetometer has as its sensor a container of a proton-rich fluid encircled by wire windings that carry a current (Fig. 3.10). The current, 1–1.5 amperes generally, can generate a locally brief, yet strong magnetic field that "polarizes" the protons. After the field is turned off the protons precess together until their motion decays into random modes about the field lines (of magnetic force) of the geomagnetic field (Fig. 3.10). There is a known frequency for this precession about the geomagnetic field that is termed the Larmor Frequency; using an average value of the earth's field of 50,000 nT it is equal to 2,100 Hz. The protons precessing in unison at or near the Larmor frequency induce a weak electrical current in the sensor's windings which, when amplified, converts the voltage's frequency to a direct measure of the magnetic field strength to about 0.02 Hz or 0.5 nT (Leute 1987).

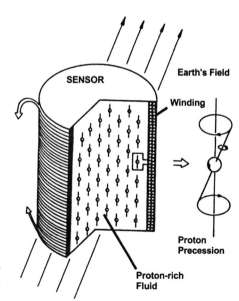

Fig. 3.10. Drawing of a typical proton-precession magnetometer sensor. Introduction of a current (~1 ampere) through the windings creates a strong, temporary magnetic field, H, which polarizes the protons in the enclosed fluid

3.8.3
Optically Pumped Magnetometer

In the geomagnetic field electronic levels will have their orbital "degeneracy" removed so that a discrete energy of a few electron volts exists between the + and – energy states of an atomic electron. This effect was discovered by Pieter Zeeman in the nineteenth century and named for him. Use of the Zeeman effect created a third type of magnetometer called the optical-pumped or absorption cell magnetometer. The electrons of a specific element such as sodium, cesium or rubidium, can move between the Zeeman levels, generally from ground (lowest) state to higher energy states by absorption of appropriately polarized light protons (hv). Optical or absorption cells filled with a vapor of sodium, for example, will experience "pumping" of ground state electrons into higher Zeeman energy states until the ground state is depleted. Absorption stops as the cell becomes opaque. Absorption is re-initiated by applying a weak high frequency radio field to the cell and the higher electron energy states will "de-populate" and cascade back down to the ground state. Optically pumped magnetometers have been in use almost as long as proton-precession and fluxgate models. A recent use of the Zeeman effect in the proton precession design takes advantage of a resonant proton-election coupling in an organo-electrolyte (paramagnetic) that increases the magnetometer's sensitivity to 0.01 nT. The effect is termed the Overhauser effect and the device based on this effect is the Overhauser-Abragam magnetometer (Milson 1996).

3.9
Gradiometers

Almost any type of magnetometer – proton, fluxgate, optical – can be configured to function as a gradiometer (Fig. 3.11). Two sensors are paired at a fixed distance apart such that a horizontal or vertical *gradient* exists between the pair. This magnetic gradient or separation is small, 0.5–1.0 m, compared to most magnetic features. Gradiometers react to near-surface anomalies with the lower sensor, if the instrument's sensors are aligned in a vertical mode, detecting the buried feature more strongly than the upper sensor. The upper sensor can be thought to be affected more by the magnetic field of the earth rather than the feature. By subtracting the readings of the paired sensors the gradient is obtained. In modern instruments this is done electronically and digitally displayed. To get a similar result, two magnetometers are often used with one remaining at a fixed location – the base station – and a second unit – the mobile unit – used to conduct the survey. At each location the readings, taken simultaneously, are subtracted to give what is termed the *differential* magnetic reading. This differential reading subtracts the magnetic field reading of the base station from the mobile reading. It is similar, but not quite the same as the gradiometer reading.

Fig. 3.11. Gradiometer.
(Photograph by author)

3.10
Magnetic Anomaly Interpretation – Basic Principles

Magnetometers measure the field that results from the sum of the magnetic field strength induced by the earth's field (average value = 50,000 nT) and any additional permanent magnetism or remanence. The remnant field component of the total field, at the measurement point, is often off-axis to that of the earth's field. Magnetic features are termed "anomalies" in the parlance of geophysics. Unless the anomaly has a strong remnant component, such as a hematite ore body or a buried clay hearth or kiln, most features found by archaeological prospection are the result of magnetic induction due to higher than average amounts of ferri-and-ferromagnetic minerals and their increased magnetic susceptibility. Magnetic prospection relies on the contrast in terms of magnetic susceptibility among buried materials that is expressed as k and total magnetism is M. Less rigorous treatments reduce to semi-empirical characterizations of the anomaly's strength such as:

$$T = kM/r^n \cdot \text{dimension}/\text{shape factor}$$

Here, T is the anomaly strength, r the distance to the anomaly from the sensor, the exponent n, is the rate of decay of magnetic strength or "*falloff factor*", and a shape or dimension factor, M and k are as before. This expression is reduced to:

$$T = M/r^n$$

Field inclination is indirectly considered by its effect on total magnetic intensity such that: $T = 2M/r^3$ for a dipolar anomaly along its axis; $T = M/r^3$ for a dipolar anomaly off axis (Fig. 3.12); and $T = M/r^2$ for a monopolar anomaly, any orientation (Fig. 3.12). Field measurement of I is difficult and is most commonly estimated, e.g., 60° inclination at 30°N latitude. If I is known or estimated, a simple vectorial solution for the anomaly strength can be obtained.

The falloff factor is important. The field of a dipole falls off as the third power of the distance. For a monopole this is only by the square of the distance. In practical terms this implies that the strength of an anomaly, measured at 81 nT directly over the feature, will fall to 3 nT at a distance of 1 m. By way of comparison, active EM devices such metal detectors have a total falloff factor of r^6.

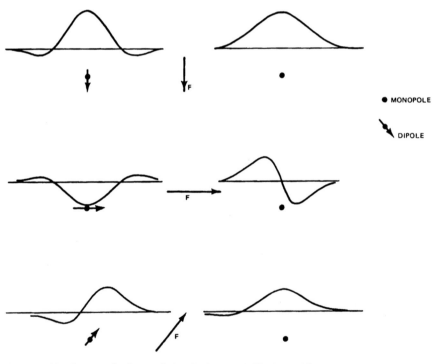

Fig. 3.12. Simple anomaly shapes. (After Breiner 1973, Fig. 25, p. 22)

Even if assuming a purely vertical or horizontal field direction, one can ascertain much about the nature of an anomaly by observing the rate of falloff and the sign of the anomaly along with its strength. Anomaly shapes change with an increase/decrease in sensor height and field direction. Depth estimation is likewise important. One can use the straightforward relationship seen in the above formulae or use "rules of thumb" or nomograms. Milson estimates the accuracy of these latter methods to within 30% (Milson 1996). One method is the "full-width half-max" rule (FWHM). The anomaly's maximum value for Z (field strength) is used to determine those points on the x axis where the value falls to Z/2. The distance between the two points of Z/2 is proportional to the approximate depth of the anomaly. For a spherical dipole, the depth $\cong x/2$ and for a monopole the depth $\cong 1.3x/2$. In the mid-latitudes of the northern hemisphere the maximum value of an anomaly tends to occur to the south of the actual feature at a distance equal to one-third the depth of the feature.

3.11
Magnetic Prospection – Practice

Many "successful" magnetic surveys have been done in archaeology without a great deal of attention paid to first principles. In those surveys the results were judged on the presence or absence together with strength of anomalies. Indeed, with the first magnetometer models this was about all one could expect before reliable digital measurement of the field strength was added. Martin Packard and Russell Varian are credited with the development of the proton precession instrument in 1954 (Weymouth 1996). By 1956 John Belshé demonstrated that a replica of an ancient kiln produced a magnetic moment. Martin Aitken and Teddy Hall of Oxford constructed a proton-based unit and located an actual kiln buried in a Roman site (Aitken 1974). Linnington at the Lerici Foundation in Rome surveyed the Greek site of Metapontum in 1968 and began the modeling of magnetic data. By the 1960s, workers such as Scollar in Bonn were doing automatic data collection and applying computers to the processing and display of that data. Workers such as Aitken, Linnington, Tite, Scollar, Ralph, Beven, Weymouth, and others rapidly moved magnetic prospection forward in archaeology.

Understanding the susceptibility contrasts of materials greatly enhances the predictability of what one can hope to detect in differing field situations. For instance, a contrast of 10^{-2} vs. one of $\Delta k = 10^{-4}$ results in strikingly different anomaly strengths, i.e., 32 vs. 0.32 nT. Another factor not mentioned in the discussion of principles is the daily or *diurnal variation* in the earth's field. This is a real and fairly regular effect observed over the course of the day. This change in observed field strength is a result of ionospheric currents that grow in strength over the day with solar heating and subside overnight. The pattern in mid-latitude areas follows an increase in field strength to 11 am, a decrease of 10 or more nanoteslas towards dusk, then a steady increase up to midnight.

Again the increase is in the low terms of nT. Long duration surveys must account for the diurnal variation. This is typically done by using the differential arrangement discussed previously. These changes are subtracted from the data to normalize the a.m. and p.m. components. If a gradiometer instrument is being used then such a correction is less necessary. Another common factor that intrudes on long traverses (>100 m) is the earth's field gradient. This is a consequence of running most profiles in a N–S direction. Those readings most northward will gradually increase as a function of the normal increase in the earth's field strength as one approaches the poles. It is easily observed as a slight change of profile slope and subtracted from the data.

3.12
Archaeological Application of Magnetic Prospection

Example. Mission Santa Catalina de Guale, St. Catherine's Island, Georgia (USA).

This site, previously discussed in regard to resistivity survey techniques, was re-discovered using a proton-precession magnetometer in 1981. In coastal island and plain surveys of the southeastern US coast, the sandy soils there are relatively iron-depauperate, due to rapid weathering and low amounts of iron minerals. Pre-European contact archaeological sites are likewise metal-poor with only the rare native metal found in burials. Even on early colonial sites of the sixteenth and seventeenth centuries metal items, if of a ferrous nature, are quickly oxidized and their original magnetism greatly reduced. The site consisted of badly decayed shell-tabby walls which may or may not have been destroyed by fire (Thomas 1987; Fig. 3.2). These walls were covered by sandy sediments 0.25–0.5 m in thickness. The effect of diurnal variation of the earth's magnetic field was evident in the survey of the site as the first morning produced discrete monopolar anomalies over the line of the walls of just over 10 nT. The anomalies (walls and well) were near the beginning of the survey traverses so gradient increases were luckily minimized. Had the detection of the walls occurred toward the end of the long traverses late in the day, the reduction of 4–6 nT in anomaly strength could have made their detection difficult because of the natural effect of diurnal variation. As it turned out a re-survey of the site in the early morning the following day demonstrated the wisdom of locating low amplitude anomalies during periods of low diurnal variation. The magnitude of the anomalies were reduced over half to only 4 nT on average.

3.13
Data Acquisition and Display

Magnetometer surveys, as with most of the other techniques discussed in this chapter, are done on orthogonal grids of straight headings and fixed offsets between (traverse) lines. Other than sensor height these are the only real variables in a magnetic survey that an investigator can control. As just discussed geomagnetic and ionospheric factors are accounted for, but not chosen beforehand. Unlike other geophysical surveys, archaeological prospection involves more grid-area survey on closely spaced (0.5–2 m+) lines. The spacing of grid intervals is largely determined by the size of the feature(s) being sought and susceptibility contrasts. For small, discrete objects of low-to-moderate magnetic contrast the grid should rarely exceed 1 m in interval/offset spacing. For larger features such as structural remains, roadways, etc., the interval can be relaxed to 5- and even 10-m spacing if the area to be surveyed is large.

On average, one team of two to three individuals can survey a 20 × 20 m area on a 1-m interval in a 6–8-h period. Speed in these surveys is determined by the cycling time of the instrument which is basically a nonfactor with any modern instrument, given that the operator can occupy a point, measure, move and repeat the sequence in a typical one-half-second cycling time of a precession device. Optically pumped systems read continuously so one just walks, reads, etc. This aspect of magnetic survey is its great advantage. Even compared to the two-probe resistivity array magnetic surveys in terms of time make it a tortoise-hare situation. This time the hare wins. Other comparable techniques, in terms of speed, are some EM survey devices and in some instances GPR.

Data produced by magnetic prospection are three-dimensional, i.e., X, Y, Z or "triplet-form." The Z value is that of the magnetic field. Because mapping software today commonly uses triplet-form data, it is easy to substitute magnetic field strength for altitude producing both (1) contour and (2) isometric plots of the data.

Example. Scull Shoals village (Georgia) smithy, magnetism and conductivity of buried ruins.

Figure 3.13 displays magnetic data of a ruined smithy in the archaeological site of Scull Shoals. This abandoned industrial village (cf. Fig. 3.8) had its heyday in the nineteenth century when hydraulic power was the mainstay of the US industrial revolution. By the early twentieth century, an increased frequency of flooding by the nearby river led to the town's abandonment and its return to upland forest. The small ruins – partial foundations and buried debris – are typical of structural remains found at almost any industrial site from the Bronze Age to modern times. As such the display clearly shows the utility of geophysics in their location and characterization. These two common forms of data display are shown in Fig. 3.13. In the first, magnetic data are displayed contoured on a 5-nT interval, then in profile. It is perhaps easier to assess individual anomaly shapes in profiles. Several spreadsheet programs do excellent profile plots. One of the most common is EXCEL. Profiles can be used for depth estimation of an anomaly (see preceding discussion). SURFER, a program by Golden Software, is commonly used to display magnetic data.

UniMag contour map with cross-section of major anomaly

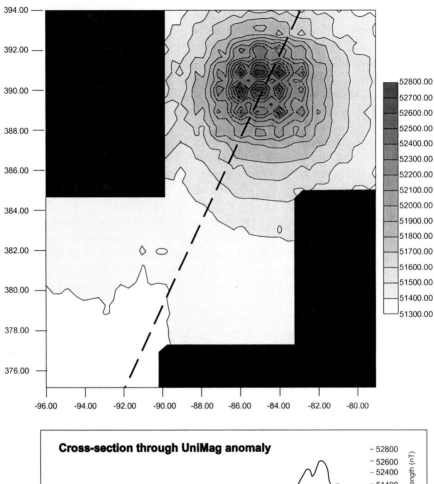

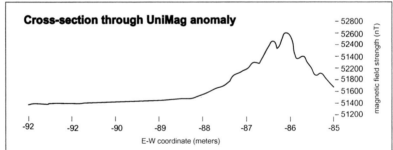

Fig. 3.13. Typical contour (**above**) and profile plots of magnetic data

3.14
Advantages and Disadvantages

Speed of survey, relative ease operation, rapid computer-based display and manipulation of large data sets have made magnetic prospection one of today's most widely used geophysical techniques. The physical principles are well understood and on the depth of research into magnetism at both the micro and macro levels. The cost of instruments has become less of a problem with the increased volume of devices available.

Of the most prominent types of magnetometers-proton-precession, optically pumped vapor cells, and fluxgate devices – only the latter are relatively immune to external noise and strong local field gradients. Even these devices may have problems in some volcanic terrains. With the exception of the fluxgate systems, most magnetometers cannot be used too close to power transmission lines or large concentrations of iron and steel. A problem, but not one of the magnetometer itself, is the interference caused by modern metal commonly found on more ancient archaeological sites. In these cases, the use of EM devices, in conjunction with magnetometers, has merit. Neither surface nor near-surface metal debris or strong field gradients significantly impact resistivity survey techniques.

3.15
Ground Penetrating Radar

The publication of a textbook (Goodman and Conyers 1998) exclusively focusing on the use of fairly recently developed geophysical technique in archaeology highlights the rapid incorporation of ground penetrating radar (GPR) into archaeological prospection. At the end of the century, after only slightly over two decades, GPR has risen to a level of recognition among archaeologists previously reserved for magnetic or resistivity instruments. This newfound popularity of GPR in archaeology, in environmental studies, geotechnical and construction areas of engineering, as well as geology, has led to new and older journals alike devoting entire issues to the applications and methods of the technique (*Geophysics* 1986).

Principles. GPR is an electromagnetic geophysical technique where a radar beam (wave) is transmitted into the subsurface and depending on different electrical properties of materials in the subsurface that beam can be diffracted and reflected at objects and interfaces (Leute 1987) (Fig. 3.14). The strength and the travel time (tt) the time delay between transmitted and returning pulses of radar reflections, can be plotted by recorders and converted to depth and dimension. GPR "sees" through freshwater, ice, soil, rock, brick, and paving although highly conductive materials such as sea water, clays and metal attenuate or occlude radar propagation. The radar pulse is transmitted

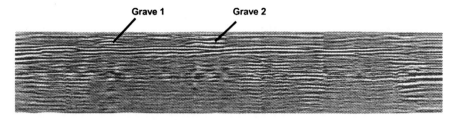

Fig. 3.14. Radargram illustrating the detection of two grave pits

as a short burst of EM energy over a compact range of frequencies called a "CHIRP". The center frequency of the antenna plus 50% above and below the center frequency makes up the chirp's range. For example, a 100-MHz center frequency includes frequencies down to 50 MHz and up to 150 MHz. The general rule in estimating GPR depth of penetration is the lower the frequency, the deeper the potential survey depth. Because GPR operates at relatively high frequencies of 20–1500 MHz, compared to seismic frequencies where "high" is considered to be in the range of 100 Hz, penetration beyond a few tens of meters in all but the most favorable, material, rock is generally uncommon. For most antennae used in archaeology, the depth range is 5 m or less, the working limit of 400–500 MHz models. For most archaeological sites this range is quite adequate as the bulk of most habitation sites seldom exceed this depth range with the exception of tells and other deeply found settlements. Even in these sites of exceptional depth the antenna with a frequency appropriate to the survey can be found, with the results then being mediated by resolution considerations.

The resolution of a specific radar frequency is a direct function of wavelength, which in turn, is a function of the velocity of the radar wave in a material. Velocity relies on the value of electrical permittivity, or more commonly termed, the dielectric "constant" (coefficient), ε. With EM techniques, it governs attenuation in both cases where induction currents extract energy from the radar pulse. The conductivity of materials generally follows an increase in the dielectric constant, but this is not so direct a relation as to have seeming contradictions where fresh- and seawater share the same value ($\varepsilon = 80$), yet have extreme differences in conductivity ($\sigma = 0.5$ vs. 3000). This contradiction is better understood when the electrolytic properties – ionic vs. nonionic – are considered. Permittivity is a direct reflection of the molecule's ability to polarize. The typical values of ε, σ and velocity (v) as well as attenuation factors are shown in Table 3.3.

The wavelength and velocity are given by the following:

$$\upsilon = 1/\sqrt{\mu\varepsilon}$$
$$\lambda = \upsilon/f\sqrt{\varepsilon} \quad \text{and} \quad v = c/\sqrt{\varepsilon\cdot\mu}$$
$$\lambda = \upsilon/f$$

Table 3.3. Typical values of radar parameters of common materials

Material	ε	σ (mS m⁻¹)	v (m ns⁻¹)	χ
Air	1	0	0.30	0
Ice	3–4	0.01	0.16	0.01
Fresh water	80	0.5	0.033	0.1
Salt water	80	3000	0.01	1000
Dry sand	3–5	0.01	0.15	0.01
Wet sand	20–30	0.01–1	0.06	0.03–0.3
Shales and clays	5–20	1–1000	0.08	1–100
Silts	5–30	1–100	0.07	1–100
Limestone	4–8	0.5–2.0	0.12	0.4–1
Granite	4–6	0.01–1	0.13	0.01–1
(Dry) salt	5–6	0.01–1	0.13	0.01–1

where c is the velocity of light in free space (0.30 m ns⁻¹), υ is the velocity of the EM wave in the medium, f is the frequency, ε is the dielectric constant (permittivity) and μ, relative magnetic permeability which in most materials (except metals) is close to 1. Most calculations assume it to be unity. Velocity and wavelength change with material giving radar its ability to image the subsurface. For example, in air, a 100-MHz pulse wavelength is 3 m; in rock with a velocity reduced to 0.1 m ns⁻¹, it is only 10 cm. In wet clay, this will drop to about 1–2 cm. As we have said, resolution is a function of wavelength so features in rock can be resolved to 10 cm, but no smaller as that dimension is inside the wavelength and thus, subject to refraction and blurring. In this the radar wave mimics other EM waves. Another general rule for GPR resolution is that the size of the object should be not less than one-tenth of its depth (Milson 1996). As a result, a 10-cm object cannot easily be detected at over 1 m unless the reflection, in terms of signal return, is exceptionally good, e.g., a good dielectric contrast and low attenuation.

An easy way to increase penetration and signal return is to increase power on the outgoing pulse. Increasing power (or gain) of the radar pulse can create other problems such as loss of detail in the upper areas of the trace of profile; refraction multiples from objects and secondary images such as hyperbolae and extraneous noise in the record. Many digital GPR units allow the operator to enhance gain losses in a GPR pulse somewhat selectively. The GPR trace is the radar waveform that is perturbed as it penetrates through subsurface materials so that the returning waveform's shape provides us with information about the earth's shallow near-surface. A GPR profile display is simply a series of traces in series taken over a traverse of some fixed distance using a set timing interval. The height of the GPR profile is a set time interval in nanoseconds. Typically, one estimates the average velocity for the soil at the site to be surveyed and this is used to set the time "window" or interval. Dry sand's velocity is 0.15 m ns⁻¹ so a 100 ns (tt) window represents about 9.5-m depth. Before presetting an unrealistic depth for the particular antenna type/frequency one can estimate the realistic depth of attenuation called the skin depth by the empirical rule, $35/\sigma$ = skin depth. Here, σ is the conductivity, emphasizing once again its important role in attenuation of the GPR signal.

3.16
Field Survey Methods

GPR in seismic parlance is a "zero or fixed offset" profiling technique. "Zero" because in some antennae, using the so-called bow-tie dipole, the transmitter and receiver are in the same antenna. Two antennae may be coupled in bipolar or bistatic configuration, so that one is the transmitter and the other the receiver yielding a small (1–2 m) fixed offset. These offsets are trivial compared to those used in a seismic reflection or refraction survey where offsets can be hundreds of meters across a geophone array. The antenna is either dragged or carried across the ground at a walking pace along the preset traverse. In some cases where exceptional signal-to-noise (S/N) ratio is required, the antenna may be "stepped" along the survey line at preset intervals. This allows "stacking" of the data to be done at each station along the traverse, thus increasing the S/N. Most field surveys require a minimum of two personnel – one is the control unit operator and the other pulls the antenna ("the mule"). Unless the individuals pre-grid the survey area, additional assistants can do this as the survey advances.

The grid size is important, particularly today, with varieties of spatial profile and 3-dimensional plots available on computer programs. Unless specific parameters are maintained the application of these data processing tools is obviated. For transverses, 1 m offset, has become a default interval for most archaeological GPR surveys. Adjacent traverse lines have a good probability of crossing features at or near this size within the grid. By halving the interval doubles survey time, almost all features at or above this size will be detected. The depth of survey is preset as we have discussed. If the site is shallow, 1–2 m in overall thickness, then it is important to set the time window to be below this, say 2.5–3.0 m range in ns. For dry sand, this is roughly 130–200 ns, for a velocity of 0.15 ns m^{-1}. This is not an impractical setting for antennae frequencies of 100–500 MHz.

Orthogonal grid directions can be surveyed. If time permits, it should be attempted. Linear features which the antenna may just parallel on a traverse, can be bisected by a crossing pattern survey. At the start of each traverse, no matter what direction, a GPR must be rapidly recalibrated for comparable data across the complete grid. This is routine and requires little in the way of resetting of instrumental parameters. Generally, the surface zero set and gains are all that require adjustment. Once a routine is established and environmental conditions allow, survey progress can be steady and consistent. A recent GPR survey conducted by our University of Georgia team covered over 8 linear km of traverses in a 5-day period. The basic grid area was 100 ×

100 m surveyed on a 1-m offset interval. A factor that has led to the rapid acceptance of GPR in archaeology is the immediate return on survey effort. GPR is a "WYSIWYG" technique – what you see is what you get. The GPR record is a facsimile of the subsurface immediately below the antenna along the direction of the traverse. Scrolling the data on a laptop screen or printing a strip paper chart provides the archaeologist with immediate knowledge of the site that he or she otherwise must obtain by other means – geophysical or excavation. With care and training, almost anyone can make a reasonable first hand interpretation of a field radar record. More detailed analyses are done with the aforementioned postprocessing routines back in the geophysical lab.

3.17
Digital Post-Processing of Ground Penetrating Radar Data

In the last decade much of the enhancement of the interpretability of GPR records have come through the use of digital processing similar to that developed for acoustical seismic methods. Once the GPR data are logged in this format, it can be readily manipulated to remove background noise, enhance gain in weak returns by reverse time-varying gain amplification, together with the use of other filters – finite impulse response (FIR) – as well as those to compensate for spherical effects in propagation. Advanced seismic style algorithms such as "Box car filtering", "Krieging" together with fourier transformation of frequency to time series are available which, in turn, prepares the data for deconvolution – which removes peg-leg hyperbolae and multiple reflections from the record. Migration, a common seismic technique, can be applied to profiles that are adjacent and are spaced closely enough (Meats 1996). The reader is directed to Goodman and Conyers (1997) for a good introductory review while the advanced student is directed to their bibliography and journals such as *Geophysics; Applied Geophysics, Journal of Archaeological Science, Archaeometry, Environmental and Engineering Geophysics*, and more recently, *Archaeological Prospection*.

A useful iteration of GPR data is the so-called time slice technique (cf. Goodman and Conyers 1997). This processing method combines data from adjacent profiles at fixed intervals of time such as 30, 45, 60 ns, etc. Using a protocol similar in form to GIS methodology where sequential overlays or "slices" are built up giving the viewer an overhead view through the site as visualized in the GPR data. Typical examples are those shown in Figs. 3.14 and 3.15.

3.18
Multiple Geophysical Techniques

Hesse (1999) has written succinctly about the combined use of multiple geophysical prospection techniques ("sensors") for archaeology. Another

Fig. 3.15. Drawing of typical Chumash Indian houses on oceanside cliff, Santa Cruz island, California (*above*). GPR profile of Chumash house floor, Santa Cruz Island (*below*). (Original drawing by Michael K. Ward; GPR image courtesy of Dr. Jeane Arnold)

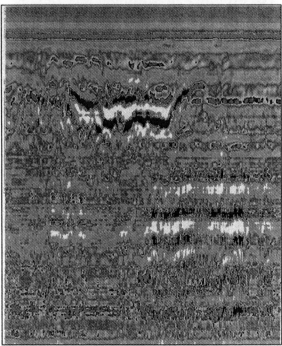

excellent example is that of the geophysical study of Roman Grand, in the Lorraine region of France, presented in a theme issue of *Les Dossiers Archeologie* (1991). The point Hesse makes is that the efficacies and short-comings of individual geophysical techniques may to a large degree offset one another in the latter case and enhance interpretability of anomalies in that of the former. Multiple sensor studies are to be attempted wherever possible. Cost and availability are prime considerations. The scale of the study is a key factor. Deploying an expensive multiple sensor on relatively small sites may be a case

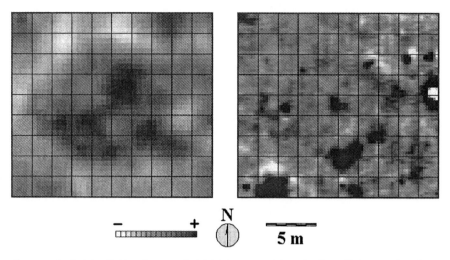

Fig. 3.16. Resistivity (*left*) and magnetic (*right*) imagery of the Whistling elk site, South Dakota, US. Resistivity plot shows the outline of the fourteenth century house with entryway; magnetic plot shows more detail, e.g. a central hearth and several posts. (Ahler et al. 2000)

of instrumental "overkill". In seeming contradiction, important, and very localized archaeological features such as tombs or burials may require the use of more than one geophysical technique to adequately relocate them in many instances. In the case of burial pits and tombs GPR has become the technique of choice (Fig. 3.14). A multiple sensor survey of the Whistling Elk prehistoric village produced excellent imagery of major structures and intrastructural features, e.g., posts, hearths, etc. (Fig. 3.16; Toom and Kvamme 2001).

3.19
Underwater Geophysical Survey Techniques in Archaeology

The use of geophysical techniques in underwater archaeology began in the 1960s (Tite 1972; Aitken 1974). These rapidly moved to assemble multiple sensor arrays as these became feasible over the subsequent decades. Today, in geophysical surveys of shipwrecks, these ensembles typically include (1) a towed marine magnetometer, (2) a 100–500 MHz sidescan sonar, and (3) an acoustic "sub-bottom" profiler, more commonly of a CHIRP, swept-frequency type. In some instances, a towed pulsed EM system is used in place of, or in conjunction with the magnetometer. Survey tracks are pre-plotted and tracked using high precision line-of-sight radar transponder systems or more and more frequently with reference to GPS. Positional accuracy in dynamic survey modes regularly approach 3–5 m in accuracy. Anomalies, acoustic, magnetic and electrical, are logged along-track to strip charts and digital files. The multiple data sets can be post-processed and reconciled just as those done terrestrially.

Two techniques that are unique to underwater survey are acoustic in nature. Deep seismic studies of the offshore for minerals have always been acoustic. Their detail, or resolution, of the sea floor and near sub-seafloor was too large scale for archaeological purposes. The late Harold Edgerton, of the Massachusetts Institute of Technology (MIT), developed a viable acoustical seismic profiler by directing a beam of a piezoelectric transducer of an appropriate frequency, 2–7 kHz, straight down to allow the beam to profile the upper sediments of the sea bottom. With sound velocities in water, ca. 1500 m s^{-1}, a 3.5–5 kHz transducer produces a waveform in lengths of 30–50 cm. Such wavelengths can regularly detect objects of archaeological interest much like GPR does on land.

The waves are pulsed or pinged by the transducer(s) activated by an applied voltage across condensers. Acoustical impedance differences caused by the relative compactness or looseness of the materials appear as returning signals or reflections of differing gray scales – dark for hard, light for unconsolidated materials. The water is transparent to the sound waves so the soil-water interface is easily seen. The returning waves, although weakened, are collected by the transducer and amplified for graphical display on a screen or recorder. Measuring the signal (reflector) and travel time (1 ms~2.7 m at 1500 m s^{-1}), the depth and thickness of a sediment reflector can be measured accurately. Like GPR the waves tend to refract and reflect more than once between the sea or lake floor and the surface generating multiple reflections. These are easily detected in the acoustical record. Individual features – rocks, buried archaeological objects – appear as diffraction hyperbolae. If the object is metal, its character can be rapidly noted by use of other techniques, magnetic or EM.

Edgerton (1986) went on to quickly design the side scan sonar after the "sub-bottom" profiler in the mid-1960s. The author was still using one of the first production types into the 1990s, producing clear sonar images of lake and sea floor. These analog print models have been supplanted by digital systems, some whose design traces back to Edgerton. The side scan sonar, by immediate reference to the output acoustical frequencies of 100–500 kHz, as are commonly used in archaeological prospection, does not attempt to penetrate the sub-seafloor. Rather, the purpose is to produce a horizontal terrain map on both sides of towed body mounting parallel transducers depressed downward at angles of 3–10°.

Figure 3.17a,b illustrates comparative sonar images produced by (a) an acoustic profiler and (b) a side scan sonar of the area of a drowned paleochannel in 20 m of ocean water on the Atlantic continental shelf 20 km off Georgia. This ancient estuary, since its discovery in 1994, has subsequently produced, in sediment cores, environmental data on the past 125–150 ka (Littman 2000).

The pace in miniaturizations of digital technology has reduced the size, and to lesser degree the cost of many of these marine geophysical instruments. The gain in performance has produced finer detail and broader scale archaeological mapping in riverine, lacustrine, and oceanic situations. At the

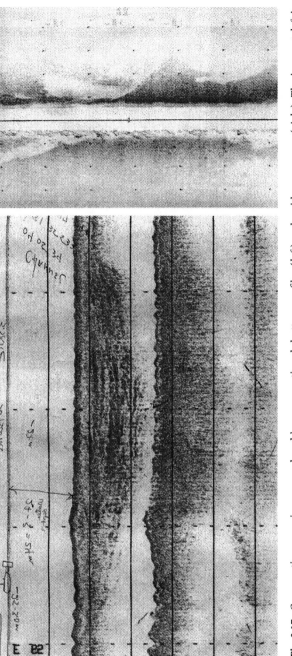

Fig. 3.17. Comparative sonar images produced by an acoustic sub-bottom profiler (*left*) and a side scan sonar (*right*). The image at *left* is that of a buried paleochannel in 20 m of ocean water 30 km off the coast of Georgia on the Atlantic continental shelf. The seafloor sonar image to the *right* is that over the paleochannel. Subsequent coring of the paleochannel produced stratified sediments dating to over 125 ka. (Littman 2000)

other end of the spectrum from lakes and rivers as well as the coastal ocean is the deep sea. This region was successfully entered and surveyed in 1975, resulting in the rediscovery of the lost ocean liner *R.M.S. Titanic*, by Robert Ballard together with scientists from Woods Hole Oceanographic Institute Exploration and France's Oceanographic Research Institute, IFREMAR. Using deep tow, long-scanning (>1 km) sonar and towed video systems the debris field and broken sections of the hull were found and identified (Ballard 1976). Using deep submarine technology, developed for the US Navy, additional discoveries were made of the lost warships Bismarck (1994); the wrecks of Iron Bottom Sound, Coral Sea (1993), and most recently (1998) shipwrecks of antiquity in the Mediterranean Basin (Ballard 1998). In the Black Sea, a proposed catastrophic inundation, proposed by William Ryan and Walter Pittman, in the Neolithic, seems to have found support from recent studies using side scan sonar, remotely operated vehicles and dredge sampling at 155 m water depths (Ballard et al. 2000).

4 Field Sampling Techniques
for Archaeological Sediments and Soils

4.1
Introduction

The focus of this chapter will be on the description of field sampling techniques for sediments and soils and their use in archaeological geology. Their use in both terrestrial and marine/lacustrine environments will be discussed along with specific applications within these environments. For the assessment of subsurface deposits and their spatial/stratigraphic extent these techniques provide the most reliable and efficient means of doing so. From the geomorphological perspective any paleosol, unless found to outcrop in the vicinity of interest, must be found using them. For ground truth purposes of geophysical data, they provide a reliable means of assessing depth of features such as subsurface contacts and electromagnetic anomalies. They provide a means to directly determine such important parameters as radar velocities in the subsurface facilitating the conversion of time to depth in ground-penetrating radar data. These field techniques allow the investigator to recover representative and depth-constrained sediment, soil and rock samples for purposes of laboratory studies.

The study of sediments is the branch of geology termed sedimentology. The study of soils is the field of pedology. "Ped" is the smallest, recognizable element of soil structure. Both sediments and soils are important dimensions of the geological study of archaeological sites. The sediment and soil context of the archaeological site is the essential matrix for the material remains of archaeological interest. An adequate scientific description of this matrix is basic to the study of any site. Without it artifacts and features lose much of their temporal and environmental points of reference. The pioneering Russian soil scientist, V.V. Dokuchiev, at the end of the nineteenth century, recognized that soils resulted from (1) the climate of a locality; (2) the nature of the parent material; (3) the amount and character of the vegetation; (4) the age of the landscape and 5) the relief of that landscape (Clarke and Beckett 1971). By a close study of this context of the archaeological site, one can retrodict many of these important factors. For example, clay minerals are sensitive to most weathering environments and reflect changes in those environments – rainfall, temperature, etc. – when measured in soils. In addition, the particle size of these minerals can help the researcher to determine the past depositional environment such as that of a back-water lake (finer clays/silts) vs. that of a fluvial nature (coarser silts/clays as well as sands).

Much of the modern study of sediments and soils is done in the laboratory and we shall discuss several of these techniques in the following chapter. Nonetheless, much information can be gained in the field. The description of a sediment or soil profile is typically done in the field as many elements of the fundamental nature of those deposits would be lost even with the most careful of sampling for later laboratory description. Beyond the macromorphological information obtained from profile studies, the collection of subsamples for later study is essential. How one goes about this involves standardized and well-known tools and procedures taught in elementary soil science, sedimentology and geomorphological courses at the college level. Still, it is important to reiterate the description of the use of these tools and techniques as this is generally omitted by editors as being too tutorial in nature.

4.2
A Brief Review of Sediments and Soils

Sediments differ from soils in the capacity of the latter to support plant growth. Many definitions of these materials focus on the organic content of a soil as compared to that of a sediment. Clastic sediment is a solid fragmental material that originates from weathering of rocks and is transported or deposited by air, water or ice. Soil is rock residuum that results from weathering over an extended period of time. To the archaeologist it is best to simply know that both sediments and soils are the result of the weathering of rock and the essential factor in the transformation of any rock to a soil is time. Except in the most exceptional cases soils require thousands of years to form in most regions of the world. Pedogenic systems differ from geologic ones in that the latter are relatively static while the former are dynamic. Pedogenic systems evolve over time and develop recognizable structural horizons, textural, biological and geochemical characteristics inherent to soils alone. Clastic sediments are sand, silt, clay, gravel, loess – particles of rocks and minerals. Like the soils from which they form, sediments are the upper surface of the earth or the *regolith*. The regolith is weathered rock and/or debris that overlies solid bedrock. In many places, such as the Piedmont of the Appalachian mountains, it forms a several meters deep cover referred to as saprolite.

Sediments and soils result from the chemical and mechanical weathering of rocks and their minerals. The breakdown of the lithic material and its subsequent redeposition is part of what Stein (1988) terms the "history of a sediment". A sediment's history is a function of four factors: (1) source; (2) transport; (3) depositional environment and (4) post-depositional environment. A study of the physical and chemical properties of a sediment or soil can evaluate the nature of these factors. By using the Principle of Uniformitarianism – the soil-forming processes active today are those of the past – and those of sedimentation one, can compare a sediment or soil of a known history and infer the processes responsible for the properties and characteristics of those under study. In archaeological geological studies of sediments and soils, the processes involved are generally both natural and anthropogenic (Butzer 1982). Once again, by examining archaeological sediments, and comparing them to reference examples from other well-studied sites, one can infer the specific processes involved in their formation (Schiffer 1987). From the standpoint of geomorphology, there are two basic processes – erosion and deposition – which determine the location and variety of sediment and soil.

As we shall see in the following chapter that a sediment's source can be evaluated by examining its textural makeup – grain-size, grain morphology and surface features, and composition (mineralogy). These parameters give insights into the transport history of the sediment with fluvial sediments demonstrating a well-known rounding due to water transport. Aeolian sediments can be recognized by grain shape and by the grain surface "polish" or glossiness. Size gives clues to the transport mechanism – large, poorly rounded gravels are good indicators of glacial transport in terrains of the northern hemisphere. Small-grained sediments are found in arid environments such as deserts and drylands. Depositional features of sediments are determined from samples either in situ or recovered en bloc. The structures of these deposits include the orientation of bedding planes, laminae of various grain sizes and textures. These depositional features are observed in profiles and cores. The last phase of a sediment's history is that of post-deposition. In archaeological contexts zonation develops either due to soil formation or cultural activities such as building, disposal of materials and the simple act of living which produces surfaces that can be identified by their micro and macro characteristics (Courty et al. 1989; Gè et al. 1993).

4.2.1
Soils

From the perspective of archaeology, soils, perhaps more so than sediments per se, provide clearly delineated markers of past landscapes and the factors responsible for their formation. Archaeological cultures exploited soils particularly after the advent of plant and animal domestication. Beyond being the important substrate for the "agricultural revolution" (Childe 1951), soils, as *paleosols*, serve as stratigraphic horizons as well as proxies for temporal/cultural periods in the archaeological record. It is more and more common to read of a past cultural facies associated with a particular paleosol, e.g., a paleolithic cultural association with a loess or tuff-derived soil; the association of the Bandkeramik culture with light, well-drained alluvial soils as described below (see also Fig. 2.13). They are equally useful to unraveling paleoclimate, glacial and other processes of interest in the study of the Quaternary.

Beginning as weathered rock such as a granite or gneiss, soils are the result of in situ formative factors: (1) climate; (2) relief or topography; (3) organic(s); (4) parent rock; and (5) time. These have been arranged into the acronym CLORPT. In the particular case of granite, a common soil is that of the ultisol. Using the parameters of the CLORPT acronym, climatic geomorphology is an area of study that has direct relevance to archaeological inquiry. While it is important to recognize that its influence on geomorphology is complex, the study of climate, in archaeological contexts, begins with the soil and its role as a proxy for paleoclimate. The location of archaeological sites has been shown to be determined by the presence of a particular soil facies. Illustrative of this are the early Neolithic sites of the Bandkeramik cultures found along the Rhine and Danubian drainages – specifically on gravel river terraces overlooking medium-sized drainages in areas of loessic soils (Howell 1987). As the Neolithic proceeds in areas such as Poland and later, Switzerland, the farming cultures move onto heavier, morainic soils (Kruk 1980).

Soils are described by use of *pedostratigraphic* nomenclature. This is in contrast to the nomenclature of lithostratigraphy which follows the NACSM code (see below). One must consider both in many descriptions of profiles as the former "overprints" the latter in all cases. Waters (1992) discusses this in detail noting that because a paleosol is superimposed onto a lithostratigraphic unit (alluvial deposit; dune; ash deposit; etc.) it is always younger than the unit on which it develops. Likewise, because the soil-forming processes alter original particle sizes and creates new pedologic structures, the original stratigraphic features such as contacts are often obscured or otherwise made difficult to interpret (Waters 1992). Soil horizons have no relationship to the geologic deposition in the lithostratigraphic units on which they occur. The description of soil horizons and subhorizons will be discussed in the next paragraphs. These are typically done in the field, but sample blocks can and are described in a laboratory setting. Certainly this is true for "deep" core samples (>1 m) obtained using devices like the Giddings-type and other mechanical coring devices (cf. Chap. 5).

Gvirtzman and Weider (2001) present an excellent discussion of a combined field and laboratory description of a 53-ka eastern Mediterranean soil-sequence stratigraphy. In a section with at least six paleosols, these researchers used a combination of descriptive protocols which can be used as a model for similar studies. In their study, pedostratigraphic, lithostratigraphic and sequence stratigraphic nomenclature are combined in a full description of the soil sequence. For the pedostratigraphy, the soil sequence is described from the *top down* while the stratigraphic sequence is described from the *bottom up* (Gvirtzman and Weider 2001). Moreover, these researchers use the terminology and methodology of modern sequence stratigraphy as applied to paleosol units such that the *sequence boundary* (cf. Chap. 2) is a surface between the top of an older buried soil or sediment and the base of a younger overlying soil or sediment. Following the standard practice in lithostratigraphic description, the paleosols were considered "lower" in stratigraphic hierarchy than "member" and were described as say a "Hamara Member" of the "Hefer Formation". Paleosols within the Hamara were described as "subunits", still maintaining a valid priority in sequence terms. The subunits are separated by the use of soil-sequence boundaries as just described. This usage of sequence-stratigraphic terminology differs from its more common application to marine sequences where "sequences" are made up of *parasequences* separated by *erosional surfaces*. In the marine context these reflect basin facies changes, whereas in a soil-sequence setting, the soils form on the "erosional" surface(s). To use sequence stratigraphy nomenclature exclusively would require that terms like "formation" be dropped. In a lithostratographic context, the contacts are the boundaries or erosional surfaces. Once again, it must be plain that one must take care in the use of these nomenclature systems and be very clear in describing their use in a particular study.

4.2.2
Horizons and Their Terminology

A solum or that upper part of the soil horizons – A and B – of a buried soil termed a *paleosol*, like a sediment, can give us information on the five factors that produced it. Paleosols are the grist of most soil studies in archaeological geology. Buried soils are characteristic of the mid-latitudes and are rare in the tropics and arctic. Russian scientists such as Dokuchiev and Sibirtsev first described soils as consisting of zones. The American C.F. Marbut elaborated on the Russian scheme giving names to differing soil types. Today, we continue the use of names for specific types of soils, but the term zone, used to distinguish soil profile divisions, has been replaced with soil horizon nomenclature.

The generalized soil profile (Fig. 4.1) consists of the "master" horizons: O, A, B, E, C, and R. O horizons are that layer above the mineral surface and consist of decomposed organic material such as leaf litter and the like. Together with the A horizon, it is also termed the "topsoil". The A horizon is the topmost mineral horizon located just below the surface. These horizons generally have a high organic content and have lost iron, aluminum and clays resulting in a concentration of siliceous material. The E horizon is now generally recognized as a separate horizon, although it is still common to see it considered by some soil scientists as a leached A or B and even a "transitional zone" between the A and B, but having no specific horizon name. E horizons are those of maximum leaching or *eluviation* of soil minerals, most notably iron, alumina and clays. The E horizon is bleached or lighter colored in appearance compared to that of adjacent horizons. The B horizon is the *illuvial* layer where the translocated mineral complexes from the A and E are concentrated. Depending upon precipitation a B horizon will vary in the amount of redeposited minerals as well as development in profile thickness. B

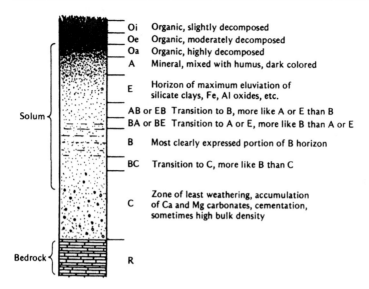

Oi   Organic, slightly decomposed
Oe   Organic, moderately decomposed
Oa   Organic, highly decomposed
A   Mineral, mixed with humus, dark colored

E   Horizon of maximum eluviation of
     silicate clays, Fe, Al oxides, etc.

AB or EB Transition to B, more like A or E than B
BA or BE Transition to A or E, more like B than A or E

B   Most clearly expressed portion of B horizon

BC   Transition to C, more like B than C

C   Zone of least weathering, accumulation
     of Ca and Mg carbonates, cementation,
     sometimes high bulk density

R

Solum

Bedrock

Fig. 4.1. Soil horizon nomenclature

horizons are mistakenly referred to as the "subsoil", by definition, making the
O and A the topsoil, no matter what their composition and thickness. The C
horizon is sometimes termed the *regolith*, but this term, as implied above, is
better reserved for unconsolidated material overlying rock (Brady 1990). The
C horizon may appear as rock, but it has undergone pedogenic processes.
Saprolytic soils are good examples of this with deeply weathered C horizons
several meters thick in some regions. The R horizon is unaltered, consolidated
country rock or parent material. The R horizon lies conformably with the
superposed sediments and soils above it.

The relationship of artifacts to soil horizons is central to any archaeological geology of a site. Cremeens and Hart (1995) provide a good review of this
important relationship in their consideration of chronostratigraphy and
pedostratigraphy, relative to the archaeological context. In that paper, the
authors note the difficulty of using the North American Code of Stratigraphic
Nomenclature (NACSN) pedostratigraphic term of "geosol", which stipulates
a paleosol must be overlain by a lithostratigraphic unit – a condition rarely
met by Holocene soils (Cremeens and Hart 1995, 15). Likewise no subdivi-
sions of a geosol are admitted; which is a problem for archaeological descriptions to say the least. Adjacent soil horizons – A, E, B, etc. – are spatially
discrete, but not temporally so. In this they differ, fundamentally, from the
lithostratigraphic description of sediments. In even Holocene contexts, paleosols can be of three types – (1) relict or a surface, never buried paleosol; (2)
exhumed or a buried paleosol now subaerially exposed and (3) buried or
"classic" paleosol. Paleosols are recognized on the basis of their organic, textural and stratigraphic features (Collinson 1996). Because of their antiquity,

the increased exposure to diagenetic processes can make original pedogenic differences between horizons more pronounced when compared to analogous modern soils. This is observed in "red bed-type" paleosol successions such as the example given for the "Hamra" soils. The authors of that study note this in their observation that "the older hamra developed (and) enhanced the red color and increased clay content" (Collinson 1996). The fluvial geomorphologist David Leigh (pers. comm.) maintains that in most areas of the eastern USA, only buried B horizons survive in paleosols. Certainly, in agricultural areas the presence of A/E horizons is generally not seen.

In the Gvirtzman and Weiner study, the soil sequence demonstrates a common aspect of paleosols which is an overprinting of a later soil onto the preceding one. This is referred to as *polygenesis* and leads to what are termed "composite" paleosols. In particular, the production of red pigments may take away from existing pedogenic features (Collinson 1996). When the soil sequence is not so mixed and the separation of paleosols is such that soil-forming episodes are distinct, then one must examine the nature of the geologic processes that "re-set" the sequence to primary stages of development. Vitousek et al. (1997), in their study of Hawaiian paleosols, suggest glaciations and volcanic eruptions are the most likely to do this. Gvirtzman and Weiner (2001) observed that coastal aeolian deposition was the mechanism at their location.

In the study of the geomorphology of the Quaternary most workers use Birkeland's cross-disciplinary approach (Birkeland 1984). For purposes of archaeology where one may be describing a paleosol or a sediment stratum, it is important to remember that the objectives of that description may be very different, as mentioned earlier in this section, from that of a pedologic or edaphic characterization. Archaeological sediments can have lithostratigraphic differences, based on texture and composition; chronostratigraphic differences, based on temporal distinctions; allostratigraphic differences, described on the basis of bounding discontinuities; as well as by pedostratigraphic differences, based on one or more differentiated soil horizons. As such, the stratigraphy of archaeological deposits can be formalized using the North American Stratigraphic Code (NACSN 1983) or, perhaps better still by the International Stratigraphic Guide (Salvidor 1994). Workers such as Stein (1993) endorse the use of cultural items such as artifacts or features to specify ethnostratigraphic differences in archaeological sediments.

Soils are classified according to systematic taxonomic naming systems. The system developed in the United States is termed the Seventh Approximation or Soil Taxonomy. This system was developed by soil scientists of the US Department of Agriculture and those of many other countries and was introduced in 1965 in the USA, and has been adopted, to varying degrees, by over 45 other countries. In this system the A and B horizons become *epipedons* or surface horizons and those horizons below the B are subsurface horizons. There are six categories in the classification system arranged in a descending order of scale – order, suborder, great group, subgroup, family and series. The

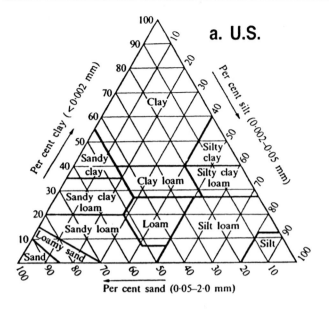

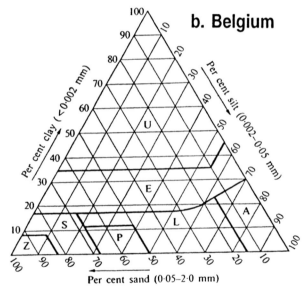

Key:

U	Zware klei	Argile lourde	(Heavy clay)
E	Klei	Argile	(Clay)
A	Leem	Limon	(Silt)
L	Zandleem	Limon sableux	(Sandy silt)
P	Licht zandleem	Limon sableux léger	(Light sandy loam)
S	Lemig zand	Sable limoneux	(Loamy sand)
Z	Zand	Sable	(Sand)

Fig. 4.2. Various textural classification schemes

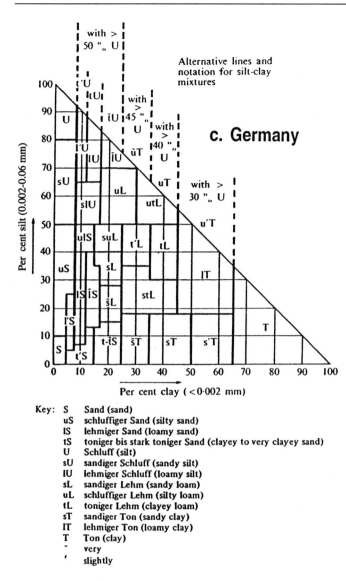

Fig. 4.2. *Continued*

Key:
S	Sand (sand)
uS	schluffiger Sand (silty sand)
lS	lehmiger Sand (loamy sand)
tS	toniger bis stark toniger Sand (clayey to very clayey sand)
U	Schluff (silt)
sU	sandiger Schluff (sandy silt)
lU	lehmiger Schluff (loamy silt)
sL	sandiger Lehm (sandy loam)
uL	schluffiger Lehm (silty loam)
tL	toniger Lehm (clayey loam)
sT	sandiger Ton (sandy clay)
lT	lehmiger Ton (loamy clay)
T	Ton (clay)
-	very
'	slightly

most inclusive category is that of *order*. There are only 11 soil orders, but there are around 16,800 soil series recognized in the United States alone.

The American Soil Taxonomic Classification System was developed to describe and map soils found on the surface of the earth today. As Fig. 4.2 illustrates, this is not the exclusive soil classification scheme. The American system; the FAO and the French schemes are the most commonly used

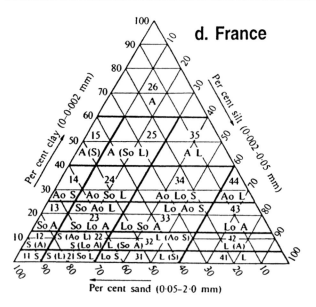

Fig. 4.2. *Continued*

(Collinson 1996). The American system relies on the organic content of the near-surface layer and the chemical characteristics of the deeper horizons. The FAO scheme, somewhat simpler, recognizes basic soil types predicated on maturity of the soil type and climate. The French system relies less on climate criteria and more on soil-forming processes.

Soil taxonomic nomenclature allows a quick and thorough description of the properties of a soil. The names of the classification units are simple combinations of syllables that are derived from Latin or Greek root terms each of which convey some key aspect of the nature of the soil, such as the Latin root *aridus* for "dry" which appears in the order name aridisol. The Latin root *mollis*, meaning "soft" forms the order term mollisol. The suffix *sol* is the Latin for "soil" so a mollisol is a "soft soil" and likewise an aridisol is a "dry (land) soil". As one descends the naming system a wet, soft soil is described by the suborder term *aquoll* which is the conjunction of "aqu(a)" and "(m)oll". Note the higher taxonomic designator – "(m)oll" appears last in the sequence. Continuing the description of the soil the level of the great group, a clay-rich, wet mollisol is a *agriaquoll*. If the soil is characteristic of a subgroup it is described as "typic" and with the addition of textural designators – "loamy", "sandy", etc. – then a particular soil could be described as a *loamy, Typic Agriaquoll*. The final level – series – is the least informative as the series name departs from the logical taxonomic pattern using, simply, a name such as "Brookston", "Westland", "Fayette", etc. to designate the soil. A buried soil or paleosol is indicated by the prefix *paleo-* to the great group designation such as a buried aquoll (a wet mollisol) termed a *Paleoudoll*. In fluvial contexts multiple paleosol sequences are very possible.

In this volume the taxonomic level of soil order is used for discussion purposes. The twelve soil orders are: (1) Entisols – soils without any distinctive layers or profile development; (2) Inceptisols – soils with poorly developed layers or weak B horizon, e.g., mineral translocation is not present or is poorly advanced; (3) Mollisols – soils with a dark A horizon or epipedon, base-rich and have strong B horizons, typical of grasslands; (4) Alfisols – soils with well-developed argillic or natric (high sodium, Na) B horizons and are well-watered, forming under deciduous forests as a rule; E horizon usually present; (5) Ultisols – soils with well-developed argillic B horizons, but low in base minerals (e.g., acid soils) typical of soils formed under native pine forests in humid areas such as the American South; E horizons can be present (6) Oxisols – soils typical of the tropics with deep, highly weathered B horizons that are very reddish in appearance ; (7) Vertisols – soils that are clay-rich, but have no B horizons and are characterized by an annual "shrink-swell" cycle producing large, deep cracks in their profiles; (8) Aridisols - "dry soils" characterized by ochric or pale profiles due to lack of organic matter; found in dry areas of the world; (9) Spodosols – soils with acidic B horizons rich in iron/aluminum oxides; A horizons are highly leached; (10) Histosols – soils characterized by high organic matter (>20%) originating in bogs, marshes and swamps; no B horizons; (11) Andisols – soils of volcanic origin with weakly developed or no real profile; dominated by silicate mineral – no clays – and with humic epipedon; previously included with Inceptisols; and (12) Gelisols have permafrost and many are cryoturbated (frost-heaving; solifluction, etc.). These soils consist of minerals or organic soil materials or both. They commonly have layers of gelic materials and histic (high organic percent) or ochric (pale) epipedons.

4.3
Describing Archaeological Sediments and Soils in Profile   9.15

4.3.1
Strata and Horizons

At a cleaned exposure or excavated profile, the first step in description is to determine the number and kind of strata or horizons present. If the deposit is not a soil, one still determines its lithostratigraphic nature. Each stratum or horizon, whichever the case, should be measured from a surface datum or

Table 4.1. Designations of master soil horizons and subordinate symbols for soils. (adapted from Olson 1981)

Master horizons		Subordinate symbols	
01	Organic undecomposed horizon	b	Buried horizon
02	Organic decomposed horizon	ca	Calcium in horizon
A1	Organic accumulation in mineral soil horizon	cs	Gypsum in horizon
A2	Leached, eluviated horizon	cn	Concretions in horizon
A3	Transition to B horizon	f	Frozen horizon
AB	Transition to B horizon; like A in upper part	g	Gleyed horizon
A and B	A2 with less than 50% occupied with spots of B	h	Humus in horizon
AC	transition horizon; not dominated by A or C	ir	Iron-rich horizon
B and A	B with less than 50% occupied with spots of A2	m	Cemented horizon
B	B horizon with accumulation of clays, iron,	p	Plowed horizon
	cations, humus; residual concentration of clays;	sa	Salt-rich horizon
	coatings (cutans); or alterations in original	si	Silica-cemented horizon
	material forming clay and structure	t	clay-rich horizon (Bt)
B1	transition horizon more like B than A	x	Fragipan horizon
B2	maximum expression of B horizon (clay-rich)		
B3	transitional horizon to C or R		
II, III, IV	lithologic discontinuities		
C	altered parent material from which A and B		
	horizons are presumed to have formed		
A″2, B′2	second soil sequence in bisequel soil		
R	Consolidated or competent bedrock		

OR measured ↺ downwards and upwards from the surface of a mineral horizon if present. These strata are sketched and characterized. If the stratum is a soil then the presence of master soil horizons should be noted. Arabic numerals, in the master horizon designation, indicate certain key characteristics such as "O2", an O horizon with accumulated organic materials. In the past (and in certain countries today) they were indicated by lowercase letters such as "Oa". Lowercase letters are still used today in the description of soil profiles such as the important designation for archaeological geologists – the lowercase "b" which indicates a buried soil or paleosol such as "Bb". The accumulation of translocated clays is commonly denoted as a "Bt" horizon. A more complete listing of these designators is shown in Table 4.1. Master horizon categories and subordinate distinctions are made depending upon the detail desired at the field site.

Lithological discontinuities or contacts are indicated by Roman numerals such as a change within an A horizon from a loessic sediment to a quartzitic sand matrix which would be designated as "IA" and "IIA". In the case of buried soils, the second or lower soil sequence is noted as follows: "A′2, B′2".

4.3.2
Color

It is important to observe and record this parameter as systematically as possible. The method most used today is the Munsell Soil Color Chart. The color chart is a collection of color chips organized in notebook form. The color chips are arranged according to hue, value and chroma. Hue is the principal spectral color. The pages of the color chart take their names from this attribute, such as "2.5YR" indicating a variation gray to black; pale red to very dusky red; reddish brown to dark reddish brown; light red to dark red. "5YR" differs in hue from white to black; pinkish white to dark reddish brown; reddish yellow to yellowish-red. There are ten hues in the Munsell system – the two already mentioned along with 5R, 7.5R, 10Y, 7.5YR, 10YR, 2.5Y, 5Y and Gley. On each page the color chips are arranged biaxially according to value (vertical axis) and chroma (horizontal axis). The value refers to the relative lightness (darkness) of the color while chroma designates the relative purity of the color (red, etc.) and its spectral strength increasing in numerical value with decreasing grayness. An example of a Munsell color descriptor for a sediment or soil could be "5YR2/1" The descriptor reads: a hue of 5YR, a value of 2 and a chroma of 1. This color is black. However, a one-unit shift in the chroma to 5YR2/2 makes this color dark reddish brown.

Recording colors using the Munsell system allows for a relative standardization in the reporting of soil/sediment color, but it is not an infallible tool. Too often the novice student is not instructed properly in the use of the color chart and what the three parameters signify. Light conditions at the time of the measurement can vary significantly in the field. Measurements made in the shadows of an excavation trench will differ from those made just outside the trench. Direct sunlight is generally avoided and many use indirect or reflected light. Colors read in the morning and evening, aside from being qualitatively darker will tend to appear more red. Moist and dry readings are taken of the sample. The simplest procedure is to take a small sample of the material on the end of a trowel and hold it behind the appropriate page of the color chart notebook. Color matches are seldom exact and the reader must make the best judgement possible, perhaps indicating the quality (poor to good) of the color match. A few archaeologists still eschew the use of the Munsell chart preferring to simply designate the color as "brownish-red", "light tan", etc. While his/her system may be readily intelligible to them individually, it lacks that quality for others. While two workers may disagree on whether a horizon is 10YR8/3 or 10YR7/4 the Munsell color designation is still "very pale brown" which in another less standardized description could be "tan", "light brown" or even "buff-colored"!

4.3.3
Texture

This parameter, along with color, is the most salient of a sediment or soil. Texture is size and proportion of the particles that make up a deposit. The four basic categories of size are: gravel, sand, silt and clay. In the United States these size categories are divided as follows: <0.002 mm: clay; 0.002–0.05 mm: silt; 0.05–2 mm: sand; >2.0 mm: gravel. The International Society of Soil

Science's scheme differs in the size range for silts and sands: 0.002–0.02 mm: silt; 0.02–2.0 mm: sand (cf. Chap. 5). Likewise, other countries vary somewhat in their size categories so it important to note which size classification scheme is reported (see Chap. 5, Fig. 5.1, this Vol.). Another size scale still in use in the USA is the *Phi* scale (Φ). This classification scheme ranges from $-8\,\Phi$ (gravel) to $+8\,\Phi$ (clay). Texture or particle size analysis is done with sieves to mechanically separate the gravel and sand grains into finer and finer divisions. Silt and clay sizes and amounts are determined by their settling rates in water.

A sandy soil consists of at least 70% sand and 15% or less clay. A silty soil contains at least 80% silt and 12% or less clay. Clay soils must contain 35% or more clay. In the field a rough approximation is done by "feel" or examining the texture by manipulating a sample of the material between the fingers – sands are gritty to the touch; silts less so; and clays display a characteristic plasticity and stickiness when wet. This latter parameter is known as *consistence* and is used to describe soils (see below). Visually, sand particles are readily apparent to the eye. Silt, in its finer particles, and clay cannot be seen without an electron microscope.

4.3.4
Structure

Aggregates or *peds* of soil, separated by planes of weakness, have four principal types: (1) spheroidal; (2) plate-like or platy; (3) prismatic and (4) blocky. Categories 1 and 2 most often occur in A horizons while prismatic and blocky peds are generally found in B horizons. A key aspect of peds are their size. Like texture there is a relative scale for the size of peds in a horizon from very fine (<1 mm) to very coarse (>10 mm) with spheroidal and platy structures. The class of spheroidal structures is subdivided into granular and crumb. Most sediment grains are not perfectly round so they will most likely fall into one or the other of the two subclasses. Blocky structures vary in size from very fine (<5 mm) to coarse (20–50 mm). Prismatic structures vary from <10 mm to 50–100 mm. As a rule the age of the soil influences the development of structure with older soils having more well-developed or *strong* peds and young soils have *weak* peds.

4.3.5
Consistence

This property of soils is measured under three conditions – dry, moist and wet. It describes a soil's resistance to mechanical stresses and manipulation. Most dry soils have little coherence and readily fall apart. Those that have not have been cemented by carbonates, iron/aluminum oxides, silica minerals other than the clays. Plinthite is an example of the cementation of sediments by iron that was translocated by groundwater. Calcium carbonates and gypsum ($CaSO_4$) produce cemented petrocalcic and petrogypsic horizons and are so-named in the master horizon nomenclature. This type of cementation is seen in aridisols. Generally moist, soil, has a coherence that varies from loose to extremely firm. Wet soil has a consistence that varies from nonsticky to very plastic; as one would expect the more plastic the material, the higher the clay content.

4.3.6
pH or Reactibility

Chemical solutions can provide indications of the acidity or alkalinity of a sediment or soil. Differential weathering of minerals translocates cations such as Ca^{2+}, Mg^{2+}, K^+, H^+, Na^+, NH_4^+ and Al^{3+} creating a great variety in the geochemistry of sediments and soils. The most common chemical parameter tested for in soils is the presence of the hydrogen ion, commonly expressed as pH. The acidic (pH < 7) or basic (pH > 7) character of a sediment or soil typically depends on the concentration of hydrogen (H^+) and aluminum (Al^{3+}) ions present. The exchange of cations and the relative amount of base formation is dependent on pH. In low pH (acidic) situations cation exchange is reduced. Base-forming cations dominate in alkaline sediments and soils. These chemical species replace the aluminum and hydrogen ions available for reactions. The chemical nature of a sediment or soil has a direct bearing on the diagenetic effects on archaeological materials such as bone, wood or other organic artefacts. For instance, bone preservation is significantly reduced in acidic soils. Precipitation, the ultimate source of water – hence hydrogen and hydroxyl (OH^-) ions – in sediments and soils, varies and mediates their acidity. Where soil moisture is low, then one should expect a low acidity as well. For example, Ultisols, formed in semi-tropical and temperate regions, should be acidic in nature.

4.3.7
Stoniness

This category was developed by British workers. The principal parameters looked for are: quantity or abundance, lithology, size and shape. A common way of reporting stoniness is as percentage of volume. This, in itself, is somewhat misleading as most determinations of percentage are done from a profile rather than a true volume. If one extracts a volume of soil or sediment it is easy to see that the size of the lithic materials have a significant influence on the measurement, which is to say that larger stones will require larger volumes. A straightforward scale for a vertical section's percentage of rocks is: slightly stony (<7%), stony (7–30%) and very stony (>30%) (Hodgson 1978). Again, large areas of a soil profile should be used to make this determination. The soils at Lithares (cf. Chap. 2) can be classified under the Soil Survey of England and Wales, as having small-to-medium stones. Under the

New Zealand system, (Hodgson 1978, 59) the Lithares soils are "very stony" (>30% stones).

4.3.8
Organic Matter

For the soil scientist, this component defines the soil from simple sediment. In the description of master soil horizons it is the key parameter in determining the O horizon. Its residence time rarely exceeds a few hundred years unless it is depositional (anerobic) environments that preserve it – inundated deposits such as wetlands and submarine landforms. In these environments microbial and biochemical activity is suppressed or absent. Organic matter has a great deal to do with soil moisture. Dark, organic soils tend to indicate greater water retention, whereas soils with lesser amounts of organic matter allow water penetration to deeper depths. The amount of organic matter in soils is relatively low being in the order of 6% or less in topsoils and even less in the subsoil (Brady 1990). Organic matter is a rich source of ions from the breakdown of plant and animal matter into organic complexes.

Outside the field setting, organic matter is categorized as *humus* – raw, mild, intimate or mechanically incorporated varieties (Clarke and Beckett 1971). In the Soil Taxonomy soils containing appreciable quantities of organic matter are described as slightly humose, humose, very humose and organic soil (Clarke and Beckett 1971, 46). It is difficult to assess these categories in the soil horizon. To the touch, organic matter can make a soil feel like silt, but the texture is different – a silt smears evenly while an organic soil sample will fragment or fray.

Histosols are soils rich in organic matter (>20%). Peats, by definition are histosols, but are more closely classified by British workers into three classes – pseudo-fibrous; fibrous and amorphous (Clarke and Beckett 1971, 72–73). Organic matter that is incorporated into the soil mass is termed *intimate* humus and can be dispersed throughout the A and B horizons giving these a darker – brown to black – color such as the typical mollisol. Organic horizons can be evidence of anthropic deposition of organic debris in cultural middens.

4.3.9
Mode

More usually used to describe the nature or distribution of minerals in a rock, it is less commonly used in describing sediments and soils. In the context of paleosols, the preservation of *primary* minerals is directly linked to time and diagenesis. When we refer to "mature" beach sands we infer that only the most resistant minerals, e.g., quartz, etc. have survived to make up the textural/ mineralogic assemblage. The common silicate minerals ranging from quartz through the feldspars to olivine respond very differently to weathering with the latter species the first to disappear, while zircon, structurally similar to olivine, persists longer in the soil (Ritter et al. 1995). By and large the heavy minerals can be used to gauge the exposure of the paleosol to normal meteoritic waters, percolating fluids such as soil acids, etc. Certainly their presence-or-absence is a barometer of the "history" of that paleosol, therefore their *mode* should be noted just as we measure the nature of the secondary minerals – clays, hydrous oxides – gibbsite, hematite, goethite, etc.

4.3.10
Boundaries

Transitions between strata or horizons are characterized as to their distinctness. In describing a soil, its mineralogy, texture, color and chemistry all contribute to the determination of horizons. In archaeological settings these attributes, plus those of cultural origin – anthropogenic – help in the delineation of stratigraphic differences. Where archaeological deposits occur as buried paleosols one must be aware of the likelihood of overprinting by subsequent soil-forming processes. Likewise, many of the changes observed in archaeological sediments may not be stratigraphic in nature. This is to say that not all *sola* in a site profile are necessarily important to its stratigraphic description. Archaeologically interesting levels of like chronostratigraphy may vary significantly in textural color and chemical characteristics. In general, stratigraphic differences are denoted by contacts, which in turn are most usually determined by changes in color and texture, together with inclusions – anthropogenic or otherwise. Micromorphological features such as lamellae may indicate archaeologically interesting stratigraphic relationships and then they may not (Anderson and Schuldenrein 1985).

Using strictly pedostratigraphic terminology the vertical boundaries between horizons (Fig. 4.3) are described as either:

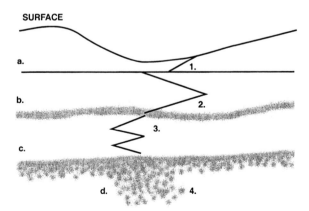

Fig. 4.3. Horizontal and vertical pedostratigraphic contacts. *a* Abrupt, *b* clear, *c* gradual, *d* diffuse. *1* Abrupt, *2* pinch out, *3* interfingered and *4* gradual

Fig. 4.4. Example of an abrupt or clear type of contact at the Ashfall paleontological site in Nebraska. The darker, upper volcanic ash overlies the lighter, alluvial sand horizon. (Photograph by the author)

a. Abrupt – if less than 3 cm wide
b. Clear –if 3–5 cm wide
c. Gradual – if 5–10 cm wide
d. Diffuse – if more than 10 cm wide

These contacts are generally defined on the basis of lithostratigraphy where breaks in the color and textural quality of the sediments are obvious (Figs. 4.3, 4.4). In geological sections, one is always cognizant of temporal gaps in a depositional sequence which are designated as conformities or unconformities. These are the boundaries between adjacent strata. Conformities, in a geological sense, have little temporal breaks in the depositional record, whereas unconformities represent temporal hiatuses where deposition has halted and even been eroded (disconformities). Conformable contacts can be described by all of the terms listed above. Unconformities are more likely to be described as abrupt, due to significant changes in the subsequent sediments. In the horizontal dimension, lateral variation in strata are described somewhat differently than in the case of their vertical relationships. These are:

1. Abrupt – the boundary between adjacent units is distinct
2. Pinchout – one unit thins progressively to extinction; not uncommon in archaeological deposits

Table 4.2. Checklist for the description of Soils

1. Group name
2. Group symbol
3. Percent of cobbles or boulders, or both (by volume)
4. Percent of gravel, sand, of fines, or all three (by dry weight)
5. Particle-size range:
   Gravel – fine, coarse
   Sand – fine, medium, coarse
6. Particle angularity: angular, subangular, subrounded, rounded
7. Particle shape: (if appropriate) flat, elongated, flat and elongated
8. Maximum particle size or dimension
9. Hardness of coarse sand and larger particles
10. Plasticity of fines: nonplastic, tow, medium, high
11. Dry strength: none, tow, medium, high, very high
12. Dilatancy: none, slow, rapid
13. Toughness: tow, medium, high
14. Color (in moist condition)
15. Odor (mention only if organic or unusual)
16. Moisture: dry, moist, or wet
17. Reaction with HCl: none, weak, or strong

For intact samples:
18. Consistency (fine-grained soils only): very soft, soft, firm, hard, very hard
19. Structure: stratified, laminated, fissured, slickensided, lensed, homogeneous
20. Cementation: weak, moderate, strong
21. Local name
22. Geologic interpretation
23. Additional comments: presence of roots or root holes, presence of mica, gypsum, etc., surface coatings on coarse-grained particles, caving or sloughing of auger hole or trench sides, difficulty in augering or excavating, etc.

3. Interfingered – the adjacent units split into smaller divisions that interdigitate with each other
4. Gradual – adjacent units blend into one another without abrupt lithological changes

Goldberg (1995), following Stein, reduces the list in Table 4.2 somewhat to the following:

1. Color (moist and dry) using the Munsell Color notation
2. Lithological composition(e.g., detrial vs. chemical)
3. Geologic vs. anthropogenic components
4. Texture (e.g., gravel, sand, etc.) and the morphology of the particles
5. Internal organization, such as bedding.
6. Consistence (e.g., loose, riable, firm, hard)
7. Voids and porosity (types and amounts)
8. Boundaries between units (e.g., abrupt, gradual, etc.)

4.4
Sampling Methods

4.4.1
Undisturbed Versus Disturbed Sampling –
Profile Samples, Soil Monoliths and Solid Cores

Undisturbed sampling of an archaeological sediment can be as straight-forward as the description of an exposed profile in the field. It can also mean the recovery of a representative fraction of those sediments for studies to be performed in a laboratory rather than in a field setting. Disturbed samples are those taken from individual horizons or strata in which structure and morphology information is sacrificed. The sense of the term "sample" used in this discussion is not one commonly associated with statistical evaluations, but simply the acquisition of a portion of the sediment or soil for analytical purposes. Sampling (statistical) procedures can, and in many cases, should be used to assure the representativeness of the field or labora-tory fraction of the sediment or soil being examined. This issue will be addressed in the following section. In this section we shall examine a variety of techniques commonly used to acquire data on archaeologically interesting sediments.

4.4.2
Profile Exposures

The simplest method of examining a profile or section is to find an outcrop or exposure. Failing this most researchers create their own by excavating one. This can be done either by hand or mechanical methods. Traditionally, shallow profile exposures are dug by hand, but in deeper sites mechanical excavation is required. At the profile face one can assess many of the primary parameters used in a description of a sediment or soil – color, textural changes, contacts, etc. The principal drawback in simple hand excavation is one of time. A longitudinal study of an excavation profile such as at an ongoing, relatively long-term archaeological excavation is the most appropri-ate way to gain an adequate field description of that profile. Such is the case in many more famous studies such as cave sites, e.g., Abri Pataud (Farrand 1975), Combe Grenal (Bordes 1972), Tabun (Jelinek et al. 1973) and Rogers Shelter (Wood and MacMillan 1975) to name some better known examples.

Mechanical excavation or exposure of a soil/sediment profile is the most common method used in geomorphology today. The most common mechan-ical means is the "back-hoe" – a tractor or tracked hoe-type excavator capable of digging to depths of 7 m with lengthy trench exposures. If one is interested in investigating fluvial valley, this type of excavator can expose entire terrace and flood plain systems within it. Unlike discontinuous sampling, typical of

coring studies, the continuous exposure by use of the mechanical hoe
provides a more synthetic view of the depositional stratigraphy – in both
vertical and horizontal dimensions.

Long-term studies of sediment profiles are the exception rather than the
rule in today's archaeology. Researchers are more often than not required to
get the information on an excavated exposure in a few days or less. Working
under such time constraints the field description of the exposure is important, but so is the recovery of a representative sample of those sediments for
more thorough study in the laboratory. Still, the field exposure provides the
most faithful picture of the exposure in terms of scale and setting. It is critical that the profile face be properly prepared before beginning its description.
Simply excavating the profile does not provide the best view of the face. One
must dress the profile face to adequately bring out the features of the deposits.
The tool of choice for this task is the archaeologist's friend – the trowel. If
archaeology has mastered one thing in stratigraphic studies, as first mandated
by Sir Mortimer Wheeler, it is the meticulous care taken with the vertical
profile. The nature of a deposit is best seen when the profile face is cleaned
and made uniformly even.

After the face is adequately prepared, measurements must be taken. The
establishment of a reference datum level is required. Most researchers
approach this step in one or two ways: (1) establish a surface datum level
or (2) establish a relative datum level on the profile face. This writer has no
preference in the selection of either method. When applied with care either
technique works equally well. In this world of "automatic" laser levels, the use
of a surface datum is manifoldly easier than in the past when workers used
string levels to transits to control verticality in an excavation. Even with the
use of the relative datum level at some time the stratigraphy must be reconciled to the actual topographic setting or geomorphic surface.

Using the relative datum level technique, the placement of the datum can
be made at any point on the profile face as long as it – the datum line – is
level. The datum line is usually a simple string level strung between two points
on either end of the profile face. From this level line the stratigraphy of the
profile can be measured up and down. The result is an accurate picture of the
profile to the scale that one has chosen. The surface datum level line is "zero"
and all readings of the profile face are below this level. The result is the same
– a representation of the profile face to scale. Where feasible a photograph of
the profile should be taken to assist in the interpretation of the profile after
the excavation has been closed. Color photography is preferred over black and
white in order to record the important parameter of soil or sediment color.
Photography of a profile face is not an easy task. Where available, a professional photographer should perform this important task. In some cases, as
with an excavation trench, the focal distance to the face is such that several
photographs must be taken to develop a composite image – mosaic – of the
profile. Lighting is almost always difficult on a profile and the best results are
obtained in complete shade rather than in direct sunlight just as in the case

of obtaining Munsell colors. Digital photography has provided a significant improvement and ease of documenting field profiles.

4.4.3
Profile Samples

Soil monoliths are small vertical columnar sections of a profile. This technique was developed by Russian soil scientists in the late nineteenth century and introduced to the rest of the world at the Chicago International Exhibition in 1893 (Vanderford 1897). The monolith can function as both a permanent record of the deposit or temporary reference section from which measurements can be made. The profile can be reconstructed from a series of these smaller sections. This procedure is laborious and can be accomplished almost as well by the quicker method of coring. The advantage of the monolith is in its use in the laboratory in assessing, under more controlled conditions than those of the field, diagnostic parameters in detail and under more convenient time constraints.

To take a monolith a container is constructed for the monolith from either metal or wood. The size of monoliths vary, but commonly are about 1 m in length, 10–20 cm wide and a few (<10 cm) cm thick. Methods vary, in detail and objectives, but the basic method is to use a wooden or metal box-like tray that is open on one side. The monolith is excavated to the dimensions of the box and cleared on all sides including the base of the section. Once this is accomplished the box is placed over the monolith and pressed firmly against it to assure a firm juncture between the container and the section. Cutting the monolith away from the profile face is perhaps the most difficult task, requiring patience and some skill with a sharp shovel or trowel to uniformly disengage the section. Some workers simply press a metal frame into the profile face and then excavate the monolith out along with the box. Either way collects a rather faithful representation of the profile being sampled. Of course, the exact nature of the deposit will condition the ease of removing the monolith, particularly in rocky sediments or archaeological deposits with many cultural inclusions. Smaller monoliths for purposes of micromorphological studies can be taken using the Kubiena frame (Kubiena 1953). These small metal boxes are again less scrupulously, but expeditiously, mimicked by the use of domestic metal food containers whose lids have been removed.

Once the monolith, of whatever dimensions, is freed from the profile it must be stabilized before transport. Thicker monoliths require less treatment before transport. The monoliths, particular the larger 1-m varieties are heavy – 25 kg or more. All specimens should be carefully logged, photographed and described before covering them with a lid. They can be safely transported to the laboratory where they are uncovered and more elaborate studies are begun. In some instances, the monoliths are impregnated with plastic resins for reasons ranging from the removal of thin sections to the fabrication of displays. From the standpoint of scientific studies such as textural, chemical and age determination, it is best to leave impregnation of the monoliths till last. One use of impregnation of soil profiles has been used to obtain "peels" of the face of a deposit (Orliac 1975). In this method a thin layer of synthetic latex is applied to the profile and allowed to harden. When dry, the latex or epoxy preserves the profile to the depth of the impregnation – a few millimeters to 1 cm or so. This image is adequate to describe color and textural differences as well as provide a facsimile of the profile that can be kept for future reference. In Switzerland, Arnold and his co-workers (Arnold and Money 1978), have used epoxy peels of entire archaeological surfaces for studies. For more elaborate particle size, chemical or other studies such as palynological ones, the peel technique is inadequate in terms of sample size as well as adulterating the sediments of the profile.

4.4.4
Core Samples *Diameter*

This type of undisturbed sediment or soil column is acquired by a variety of methods and results in specimens of varying dimensions of length and diameter. The key word in describing cores is *diameter*. The coring devices used to obtain soil and sediment samples are cylindrical as a rule. There are square coring devices such the "box corer" used in sampling marine and lacustrine sediments. Coring devices range from gravity corers, piston corers, percussion corers, drive-corers to rotary corers. The first two types, like the box corer, are used for inundated or wetland sediments. Percussion, drive- and rotary corers are typically used for terrestrial coring although a variant of the percussion type, the vibracorer, is routinely used in wetland and inundated contexts. Vibracorers are particularly effective in penetrating unconsolidated sandy sediments. A recent, rather innovative, "coring" device is the freezing corer or "cryoprobe" which has proved effective in both inundated and terrestrial settings (Garrison 1998).

4.4.5
Rotary Corers

These are mechanical coring devices that extract solid columns of sediment or soil that is relatively uncompacted. Unlike larger rotary drilling systems such as used in hydrocarbon or hydrological exploration, rotary corers used in archaeological geology do not usually use drilling fluids which are used to lubricate the drill bit in rock. As almost all archaeological deposits are sedimentological in nature, the rotary corer can penetrate their relatively shallow depths (a few meters as a rule) without lubricants, which in many instances, prove inconvenient for accurate chemical and textural tests. As shown in Fig. 4.5 the rotary corer is most often vehicle mounted for ease of transport to the archaeological site. The coring tool is an open cylinder a few centimeters in

Fig. 4.5. Two views of a mechanical auger during geoarchaeological studies at a Roman cemetery site, Carthage, in modern Tunis. (Photographs courtesy of Nina Šerman)

diameter that has a drill bit or *straight* cutting edge of hardened steel alloy. The barrel of the corer often features a screw-like exterior creating the appearance of a hollow auger that allows the central section to remain open for the acquisition of a continuous, solid sediment/soil column in all but the stoniest or wet conditions.

The advantage of the rotary corer, or for that matter any corer type, is the recovery of an *undisturbed* column of sediment or soil. The recovered monolith is reflective of the macro and microscale stratigraphy without a great deal of compaction or mixing of sediment/soil structure and features such as lamellae. For the measurement of parameters such as porosity, moisture content, and consolidation of clays, solid undisturbed core samples are mandatory (Shuter and Teasdale 1989).

4.4.6
Percussion and Drive Corers

These types of coring devices are excellent for the recovery of soft or compressible sediments (Shuter and Teasdale 1989). Likewise they work well in unconsolidated sandy sediments, particularly the percussion corer. The difference between the percussion and drive corers is the application of the pressure on the coring tool. The percussion corers apply intermittent impact pressure either by a hammer or by raising and dropping the entire coring tool. The drive corer applies pressure in a constant manner usually with a hydraulic booster system. The vibracorer is a percussion type of corer that utilizes an eccentric or offset gearing system to develop a steady vibratory motion on the core barrel enabling it to "settle" into a deposit. Vibracorers can be pneumatic, hydraulic or electric powered. Percussion corers can vary in size from 1 m in length to over 10 m. Core barrel diameters can vary from 50 to 75 mm as a rule. A popular terrestrial drive coring device is the Giddings corer. This device applies hydraulic pressure to a core barrel of 75 mm diameter extracting solid columns of over 1 m in length (Fig. 4.6).

Hand-driven percussion corers are the most popular and economical of the coring types. These corers typically consist of a solid or split-core sampler barrel, a series of solid metal extensions that terminates in a swage or screw-type adapter for a slide hammer. The individual hammers the core into the ground, in set increments, determined by the length of the coring tool, to the ultimate depth of interest and then extracts the solid core. A type of coring tool called a gouge auger can be used. This "auger" is a cylindrical open device for use in soft sediments. Hand-driven corers are rarely over 25 mm in diameter and obtain relatively small samples of the sediments/soils. Compared to some screw augers of 100 mm diameter, there is a clear advantage in overall sample size for the screw auger even if the sediment is a disturbed sample. Core liners of clear plastic can be used inside all types of corers from mechanical models to hand types. For maximum protection and preservation of the in situ context of the cored deposit, these sampling devices are recommended. This is important for the recovery of fragile macrobotanical or macrofaunal materials as well as pollen remains.

An important factor in percussion corers, most notably those that are hammer-driven, is the disturbance and compaction of fine-grained sediments. This can occur in vibrating corers as well. Quantitative data are lacking on this point, but it is suspected that an increase in the driving rate, measured in blows per foot, increases the potential for disturbance and compaction in

the recovered sediment column. The author has observed this in small diameter, rigid barrel hand corers when using a 5-kg slide hammer in resistant deposits such as stiff muds or sands. If a sample requires a driving rate in excess of 25 blows per foot then one should be on guard for sample disturbance. If the corer's hammer is quite heavy – say 50 kg or so – then this is particularly true with regard to compaction of the recovered core sample.

4.4.7
Gravity ("Drop") and Piston Corers

These corers, as previously said, are more often used to core marine and lacustrine sediments. Both types are raised and then released to rely on gravity for the force required to penetrate inundated or soft wetland sediments. In the contexts that they are deployed in, they are rarely coring into soils with the

Fig. 4.7. Ewing-type piston corer

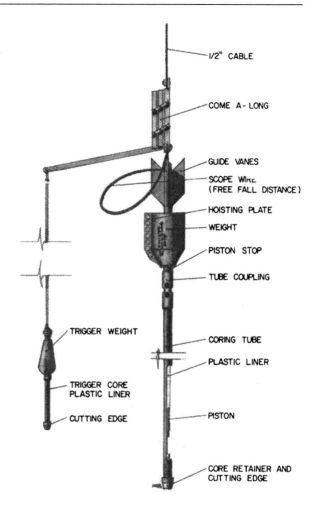

1/2" CABLE

COME A-LONG

GUIDE VANES

SCOPE WIRE
(FREE FALL DISTANCE)

HOISTING PLATE

WEIGHT

PISTON STOP

TUBE COUPLING

TRIGGER WEIGHT

CORING TUBE

PLASTIC LINER

TRIGGER CORE
PLASTIC LINER

CUTTING EDGE

PISTON

CORE RETAINER AND
CUTTING EDGE

exception of soils developed on peat bog deposits such as paleo-histosols. To aid in penetration of the sediments these corers use heavy weight loads attached to the upper part of the coring barrel with as much as a ton of weight added to corers of 2–3 m in length. Needless to say a corer of this weight is not recovered by hand.

The piston corer is a gravity corer design that uses the creation of a partial vacuum in the recovery barrel to recover wet, unconsolidated sediments. The piston corer works well in multi-bedded deposits of silts and sands. Various types are used today, such as the Livingstone corer, the Kullenberg corer and the Ewing-type. Figure 4.7 illustrates the principal components of a Ewing-type piston corer used by the author to recover riverine and marine sediments. A smaller, auxiliary corer called a messenger corer is added to a gravity corer design whose barrel assembly has incorporated a piston coupled by a cable to the messenger corer and the hoisting mechanism for both elements of the piston corer. As shown in the illustration, the messenger corer is cantilevered off the larger gravity-piston corer such that it will contact the sediments first, raising the cantilever arm such that a cam device, in turn, releases the cable at the point of attachment on the hoisting device along with the main corer. The heavier, main corer drops in free-fall mode into the sediments, while simultaneously the fixed cable remains taut along with the piston. Thus as the core barrel falls deeper into the deposits, the piston moves up the core barrel drawing sediments into the partial vacuum created by the rapid "rise" of the piston up the barrel. These corers use plastic core liners to aid in the creation of the vacuum and to preserve the sample. Piston corers are sometimes prone to problems such as foreshortening or stretching of sedimentary units (Buckley et al. 1994).

4.4.8
Freeze Corers

Hammer-driven freeze corers or "cryogenic soil probes" are a fairly recent addition to sediment sampling in archaeology. They have been used to sample wet sediments for biological purposes as well as those of archaeology (Lassau and Riethmann 1988; Grebothè et al. 1990; Miskimmin et al. 1996). Early in the 1990s European and North American workers quickly moved the design of these corers forward from prototypes to relatively standardized field models (Hochuli 1994; Garrison 1998). The principle of the freezing corer is simple – insert a closed-tip metal tube into the earth, fill it with a cryogenic fluid, such as nitrogen, and freeze the adjacent sediments to the metal tube. Extraction of the tube brings with it all the frozen sediments adhering to the core barrel. The diameters of the freezing corer barrel have varied with development as have the freezing fluids used in these designs. The present-day archaeological designs have opted for 75 mm diameter steel barrels about the length of rotary and drive corers such as the Giddings-type, e.g., 1 m or so in length. Those used in lacustrine settings are cylindrical or wedge-shaped containers of slurries of alcohol and dry ice (Lotter et al. 1997).

Figure 4.8 illustrates the design and use of a typical freezing corer. The corer is hammered into the sediments and then a transfer hose is attached to valving on the top of the corer. Liquid nitrogen, at a temperature of –197 °C, is circulated through the corer for a period of about 20 min. A mechanical extraction assembly composed of a tripod and lifting device are attached to the frozen core sample and it is withdrawn. The description of the recovered sample is described much the same as any sediment or soil sample – color, texture, structures, etc. Moisture content is important for the efficient use of the freezing corer, but it is surprising the relatively small amount of moisture needed for adequate freezing in most sediments (5–10%).

Fig. 4.8. Cryoprobe system showing nitrogen dewar, transfer hose, extraction tripod and probe tube in the earth. *Inset* shows extracted core and digital thermometer. (Photographs by author)

4.4.9
Augers and Drills: Undisturbed Sampling Techniques

Rotary augers and drills have a long history in soil and sediment sampling. Recently, two noted geoarchaeologists have discussed the relative merits of small diameter corers (25 mm) vs. larger diameter (100 mm) soil augers (Stein 1986, 1991; Schuldenrein 1991). Both types are commonly used today and their use will continue for both practical (speed, costs) and scientific reasons. Rotary augers and drills recover disturbed samples in terms of pedologic

structure and micromorphology. If the samples are in soil horizons the stratigraphic integrity of the deposits are degraded or destroyed, but much information is retained for both field and laboratory studies. Larger sample sizes
are routinely recovered using soil augers and mechanical drill systems.
Mechanical drilling systems can recover sediments as effectively as comparable solid column corers.

4.5
Handling and Description of Cores –
Some General Considerations

Depending on the type of core sample one has – auger, drill, etc. – some
general observations should be made about the sample as quickly as possible, either in the field or immediately upon arrival at a laboratory. All samples
allow the determination of the basic parameters of a sediment or soil – color,
general texture, consistency, general organic content, etc. Some things must
wait for more rigorous measures to be applied to the sample, such as the quantitative determination of size classes; total organic matter content; macro- and
micro-botanical makeup; chemical attributes; magnetic properties; etc.
Nonetheless, a researcher must recognize that the freshly recovered sample
contains information that can be readily assessed. These data, however qualitative, in some cases provide the investigator with firsthand information as
to the general nature of the sediment or soil. Additionally, these data are not
to be replicated as the simple passage of time subtly alters the core sample
through changes in moisture content and chemical changes such as oxidation
and the corrosion of mineralogic or organic remains. Rapid storage of the
core sample, in a cold room or freezer, can do much to ameliorate or retard
these changes, but even in these cases workers typically record as much data
about the sample as possible before placing the core in storage. If the core
sample cannot be archived in such facilities, then it is imperative that the
material be sampled and described as quickly and as fully as possible.

4.5.1
Rotary, Percussion/Drive and Piston Cores

These devices can extract whole core samples such that the structure of the
peds and the lithostratigraphy are generally preserved. The sample can be
extruded from the core barrel into a sample holder, or the sample can be contained within a core tube liner made of a plastic material. The core sample
can, in some cases, be maintained in a type of core barrel that can be removed
or detached to allow transport, storage and examination of the sample. These
types of cores, together with those in liners, are often "split" into two halves
with the barrel/liner performing the role of sample holder.

Depending on the length and diameter of the core sample, analysis and
storage considerations can vary. In the case of small diameter (25 mm/1 in.),

and shortish (30 cm/12 in.) core samples, handling and storage protocols are not very complex. In the case of larger diameter (>75 mm/3 in.), several meter-long cores, then handling, sampling and storage becomes a more serious issue. With the larger core samples it is not uncommon to section or cut the sample into shorter, more manageable lengths such as 50 or 100 cm. Again, in the case where the liner or barrel is used for handling, the core may be sampled and then archived without too much disturbance. Those cores that are extruded from the recovery tube immediately are altered such that the peds are adulterated in terms of shape, size, etc. Textural and color differences are maintained. The stratigraphic nature of the sample is further altered by the extraction of these types of samples. The general nature of lithostratigraphic relationships in the core can still be seen, but the absolute, metric nature of the relationships are compromised.

Whatever the core type the in-field and laboratory description of the sediment section follows a common-sense procedure in most cases. A standardized form, such as the example shown in Fig. 4.9 is used. The categories used in the core description can vary and be easily customized to the particular objectives of a project. The categories shown in these examples were used in the description of marine vibracore samples, but they could as easily be used for cores of a terrestrial fluvial system. Our example cores were contained in 3 in. (7.5 cm) diameter aluminum pipes within which they were recovered. Upon recovery they were "split" into equal halves using power saws. The halves were measured and described as to color and sediment texture. They were drawn on standardized recording sheets, as shown, and photographed along their entire sections. From one half of each core sediment samples were drawn at set intervals (ca. 20–30 cm). Depending on the lengths of the cores, typically 3 m, this resulted in about a dozen sediment samples of 100–300 g each. These sediment samples were placed in plastic bags, marked and catalogued as to location and time of recovery as well as depth within the individual core. The other halves of the cores were covered with mylar and aluminum wrap for their entire lengths. The sampled halves will be treated the same way as well. Both sets of cores – sampled and unsampled – were transported to laboratories for further study and subsequent storage in controlled environmental conditions and reserved for future research. Standard drill core boxes, made of heavy corrugated Kraft cardboard or corrugated plastic are available from most geological/soil science supply houses. One box can hold up to 3 m (10 ft.) of core sample.

4.5.2
Rotary Augers and Drills

The core samples from these devices have, by their very nature, lost pedological and lithostratigraphic information. The use of these devices are typically limited to shallow sediment/soil sampling (<2–3 m), but as most archaeological deposits are within this range, their utility will remain high.

9.27

Meter	Graphic Lith.	Section	Age	Structure	Disturb.	Sample	Color	Description
1								
2								
3								
4								
5								
6								
7								
8								

Fig. 4.9. Example of recording form for core samples

The hand augers and drills allow very rapid sampling for small, yet representative samples. Color, textural or even chemical analyses require only a few grams of sample in most cases so the volume of sample needed is not that great. Again, it is the continuity and cohesiveness of the sample that are at issue and if these are not a methodological priority then the use of the devices is feasible.

Like the more intact samples from the rotary/push/percussion corers, handling, analysis and storage depends a lot on the sample size. Long, continuous sediment samples may be extracted using the hand drill or auger. Mechanized units certainly can do this. The length of the sample is only limited by the number of extensions that are used in the recovery of the core. Samples from the core may be placed in drill core boxes or simply placed in plastic bags. These latter items are labeled as to the location and weight of the sample as a well as the other identifying information – core sample number, date, etc. One problem encountered by this author in the use of a mechanized drill was the difficulty in monitoring and recovering the sediment sample during the course of the coring operation. This particular unit did not have a variable speed control on the drill so the sample simply extruded from the bore hole as the drill descended. This type of drill is to be avoided if controlled sampling is a goal of the study.

4.5.3
Cryogenic or Freeze Cores

While relatively rarely used in coring studies today, more and more references to the use of these devices have appeared in a wide variety of literature, mostly to do with the sampling of wetland and/or inundated deposits. The author's use of these devices has demonstrated some handling problems that are unique to them. One major difference in the core taken by a freeze corer vs. that recovered by other types is temperature. Whether the coolant used is carbon dioxide($-78\,°C$) or colder nitrogen ($-197\,°C$), the recovered core remains frozen to the core barrel for some time – up to over $1\,h$ in many instances. The core sample is rock hard until it finally melts and even then the sample rarely loses its consistency and cohesion. This is not a bad thing. Indeed, it is one of the real advantages of the freeze corer that relatively uncompressed samples can be recovered intact over considerable lengths ($>1\,m/core$). Too often, in inundated sediments, even with standard coring devices with sample catchers in place, some or all of the sample has been lost during recovery of the core sample. Freezing cores do not suffer from this shortcoming.

Another interesting aspect of a frozen core is the necessity, upon occasion, to literally defrost the sample. This is done using a portable propane torch. The objective is not so much to thaw the core, but to remove the frost that forms on the surface of the sample obscuring the color of the material. Once thawed the battery of analytical techniques that can be applied to the freeze core are no different from those used on other core samples. The removal of the sample requires the cutting away of the core from the barrel to which it adheres. This can be easily done with a trowel where the length of the core is sectioned to the barrel and the sediments simply peeled back along the length. The sample can be transferred to core boxes or other sample containers.

A final consideration should be given to the determination of the chronostratigraphy of a core sample. Subsampling of the core, for textural, chemical and botanical analyses, is relatively straightforward. The recovery of samples for use in age dating is relatively routine as well, but in certain cases special handling is required. For instance, the paleomagnetic/archaeomagnetic evaluation of a core sample can only be done on intact samples. The dating is typically done in situ along the core using a magnetometer designed for this. Most archaeological dates are not obtained from the magnetic dating of cores, unlike geological core samples where the procedure is more common. For one thing, paleomagnetism measures field reversals across a broad time range, whereas archaeomagnetic dating requires declination and inclination measurements which are compared to regional reference dating curves (Eighmy and Sternberg 1990). Radiocarbon dating of organic inclusions is the most typical means of age determination of cores within the practical range (~50,000 years) of the respective carbon dating techniques – beta counting or accelerator mass spectroscopy (Blackwell and Schwarz 1993). As optically stimulated luminescence (OSL) dating of sediments becomes more common, the need for special handling of the particular core samples will be needed. Basically, this entails the use of darkness and photographic darkroom procedures to prevent undue exposure of the core sediments, such that the remnant OSL signal is not affected (Godfrey-Smith et al. 1988).

4.6
Scale and Size – Adequacy of the Sample

Sample size is relative to the nature of the deposits being examined. One must always be cognizant that the volume of sediment recovered in the average diameter corer or auger barrel is inconsequential compared to the overall volume of most deposits being studied. Understanding this simple fact places a premium on the logical extension of a description of a small sample to the larger whole of the sediment or soil volume. A common sense approach is to obtain the largest sample possible relative to the research question being asked.

If the question is one of the correspondence of a sediment or soil type with your description, then a statistical approach to establishing commonality or difference is appropriate. It is a question of the adequacy of a sample for determining paleoenvironmental data from soil/sediment structure; geochemical properties; pollen and other micro-botanical or micro-faunal inclusions such as phytoliths or foraminifera. Understanding the relationship of soil or sediment properties with respect to sample size is essential to the design of effective sampling strategies. In a paper by Starr et al. (1995) the effect of sample size on common soil properties – bulk, density, moisture content, pH, NO_3, P – was measured across five different corer diameters (17–54 mm) and a 20×30 cm monolith. The findings indicated that smaller diameter samples gave sample means greater skewness and higher variances than the larger block sample (Starr et al. 1995). Replicates or several repeated samples using the smaller corers could compensate for the variation in the data. This is by simply increasing the sample size. With an eye to the debate between Stein and Schuldenrein concerning corers vs. augers, at least with hand-driven types, the larger diameter augers (100 mm) are less likely to be prone to statistical variability.

Still the problem of arguing from the sample level to that of the population, as a whole, remains both a question of sample size and *representativeness*. The latter property relates directly to the location of the sampling. This is less a size problem than it is a spatial one. In statistical sampling theory, the issue of randomness of the samples is a key one. Many samples are taken on grids set across vertical profiles or across horizontal plots using a systematic approach – that is, every point on the grid is sampled. This is much the same as point counting procedures of grains of sediments or soils. It is practical, but nonprobabilistic in nature. The author has used both simple random and stratified random sampling where sampling locations are assigned by use of randomization such as using a random number generator (calculator) or a table of random numbers. This has been done in samples from soil horizons as well as across archaeological site grids. The effect of random sampling is to improve the representativeness of the sample and to reduce the overall sample size. Without delving into statistical models at this point, it is sufficient to say that one must be aware of *what* one is sampling.

Measuring correspondence between sample populations is generally done by bivariate comparisons of soil or sediment properties. With multiple samples or replicates one may then resort to multivariate methods which will allow the evaluation of likeness or difference in the sample populations. In the rubric of statistics one measures the parameters vs. the specific cases or vice versa. Those samples with similar variability which meet statistical tests of similarity such as chi square, t-test, etc. can be reasonably assumed to represent the same or similar populations (Hawkes 1995). Underlying these tests is the assumption that most parent and sample populations are *normally* distributed – that the average or mean of the population of sand grains – grain size – lies near the center of the size distribution and all other sizes lie to either side of the mean creating the familiar bell-shaped curve or frequency distribution. If all is well, and your distribution is normal, then sampling measures will work. If not. . . .

5 Analytical Techniques for Archaeological Sediments

5.1 Introduction

Jean-Claude Gall (1983), in his volume *Ancient Sedimentary Environments and Habitats of Living Organisms,* recognized the kinship of archaeology and geology in the fact that both face a "fossilized world". If we are interested in the paleoenvironment of that world, then the only recourse is "through the organisms and the sediments" (Gall 1983). In the case of the archaeologist, one must add cultural (artifactual) remains as well. The study of archaeological sediments and what they contain as "fossils" – organic remains such as floral and faunal materials; artifacts and nonportable cultural features, textural and chemical indices. These can provide a great deal of evidence about a paleoenvironment within which a past human population, together with plant/animal communities.

Most depositional environments, geological or archaeological, result in stratigraphic sequences or horizons. The paleoecological, sedimentological and artifactual information must be collected level by level and evaluated quantitatively, to provide, at best, a fragmentary synthesis of a prehistoric milieu. As Gall and others have noted, the paucity of that fragmented record, plus its inherent alienness, makes understanding elusive and speculative. Nonetheless, it is the record archaeological scientists must use and the sedimentary fraction is no small part of the picture. In geology, the study of sediments is the field of sedimentology. Because of the linkage between sedimentary basins and hydrocarbons, there has been a strong focus on this aspect of earth science both in terms of theory and practice. Because of their importance in the natural rock cycle, sedimentary rocks and their genesis – mineralogy, depositional histories and environments – have been examined to the extent that much information salient to archaeological enquiry can be deduced from the sedimentary matrix or context in the site. Today, with the development of optical stimulated luminescence (OSL) technology, and cosmogenic nuclides, ^{14}C, ^{10}Be, ^{26}Al, the "chronometric" or "numeric" dating of soils and sediments is possible (Wagner 1998).

While the possibility of directly dating an archaeological sediment facies is exciting, this only adds to the other well-established techniques used in the analysis of the sediment such as what I have called the "five P's", particle size, point counting, palynology, phytolith and phosphate procedures. As Julie Stein, Gall and others pointed out, the study of sediments allows us to deduce their "history" such as origin, transport, and deposition (Stein 1988). The specific study of a particle's mineralogy (petrography of sediment particles, particle size and point counting) is discussed in the two following chapters.

Simply studying the morphology and variety of sedimentary particles allows a rather reliable retrodiction of processes such as geological source, climate and diagenesis. If the sedimentary particles are found in structural materials – daub or brick – or ceramic artifacts, additional understanding of their sources, trade or other aspects of these materials can be studied. The study of those smallest of sediments – the clays – occupies a subdiscipline of its own and has proven quite pertinent to the study of archaeological ceramics. The mean size and percentage of sediments together with the calculation of straightforward coefficients, such as (1) grain size, (2) sorting and (3) skewness, provides significant information on transport, deposition and diagenetic processes. For instance, the nature of cultural vs. natural deposits in cave sediments can be examined by particle size, shape and distributional (mean size, etc.) as well as other micromorphology procedures (Courty et al. 1989; Courty 1992; Goldberg 1995).

Bulk density and carbonate percentage are important parameters determined in the laboratory setting. The latter technique is important in the characterization of sediments and soils where the buildup of calcium carbonate is commonplace and a key parameter in soil genesis as well as sediment deposition. Figure 5.1 is an illustration of a typical flow chart for laboratory procedures used in the study of soil and sediment samples.

5.2
Particle Size Analysis

Particle size analysis (PSA) is the measurement of the size distribution of the individual particles of a sediment or soil. It involves the destruction of dispersion of sediments into their constituent particles by chemical, mechanical or ultrasonic means and their separation into size categories or grades by sieving and sedimentation (Gee and Bauder 1986). As these particles range from rocks to clays, various methods of their classification have been developed using arbitrary size classes or limits as illustrated in Table 5.1.

After color determination, grain-size determination is the most fundamental test performed on soil/sediment samples. Three general size or textural classes are used by most investigators to show the textural divisions in a particular sample. These are plotted on ternary diagrams whose axes are 100% clay, silt and sand (Fig. 5.1). In the USA, the standard protocol is to use the United States Department of Agriculture (USDA) classification of the texture of sediments. In Europe and elsewhere, the particle size classes vary slightly from that of the USDA scheme. An in-depth reconciliation of the various classifications is described elsewhere (Hodgson 1978). The USDA ter-

✱ DIAGENESIS - THE PHYSICAL AND CHEMICAL CHANGES OCCURRING IN SEDIMENTS BETWEEN THE TIMES OF DEPOSITION AND SOLIDIFICATION.

[Handwritten margin notes: 9.28; THE "FIVE P'S" OF SEDIMENT ANALYSIS: 1) PARTICLE SIZE 2) POINT COUNTING 3) PALYNOLOGY 4) PHYTOLITH 5) PHOSPHATE PROCEDURES]

Fig. 5.1. Triangular or ternary-type
diagrams

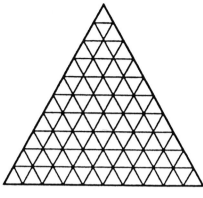

TEXURAL CLASSES

SAND
SILT
CLAY

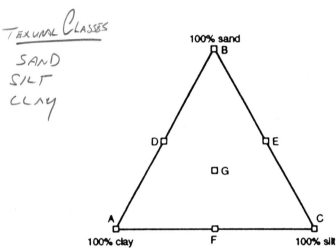

minology has gained wide acceptance outside the USA, but even in countries
such as Belgium and the Netherlands where the USDA terminology is used
for other soil features – horizons, etc. – these countries have set up their own
particle size classifications (Hodgson 1978, 54). While this heterogeneity
looks to be the case for the foreseeable future, the use of the ternary diagram
to represent the variety of particle size classes is relatively a standard
practice. Figure 4.2 showed the USDA particle size scheme (1) together with
examples from (2) Belgium, (3) France, (4) Germany, (5) Switzerland and (6)
England and Wales.

Typical PSA methods require the sediment to be broken down in to its
individual particles by both chemical and physical means. This two-part procedure is necessary in most sediments where clay minerals comprise a significant portion. The larger size particles (>2 mm) are sorted and sized by the
use of wire-mesh screen sieves in mechanical shakers (Fig. 5.2). Of course,
the sediment must be dry before either mechanical or chemical sorting is to

Table 5.1. The standard grain-size scale for clastic sediments[a]

	Name	(mm)	(μm)	φ
G	Boulder	4096		−12
R	Cobble	256		−8
A	Pebble	64		−6
V	Granule	4		−2
E	Very coarse sand	2		−1
L	Coarse sand	1		0
	Medium sand	0.5	500	1
S	Fine sand	0.25	250	2
A	Very fine sand	0.125	125	3
N	Coarse silt	0.062	62	4
D	Medium silt	0.031	31	5
	Fine silt	0.016	16	6
M	Very fine silt	0.008	8	7
U	Clay	0.004	2	8
D				

[a]As devised by Johann A. Udden and Chester K. Wentworth (Blair and McPherson 1999), the φ scale was devised to facilitate statistical manipulation of grain-size data and is commonly used. $\varphi = -\log_2 D(mm)$ where D is the grain diameter. It is a geometric scale where the succeeding size class is twice the size of the former class or size grade.

be attempted. The ease of sorting or dispersion depends on the nature of the sediments. Because of this fact, there is no one "standard" PSA technique.

Sediments rich in organics and clays are clastic and require substantial chemical pretreatment with agents such as sodium hexameta-phosphate (HMP), hydrogen peroxide, sodium hypochlorite, sodium hypobromite, and potassium permanganate. Acid treatments are generally to be avoided because these reagents, such as HCl, can destroy the lattice of clay minerals. HMP is the most commonly used dispersant chemical and is recommended in microbotanical separations of pollen and phytoliths (Yaalon 1976; I. Rovner, pers. comm.).

For small size particle aggregates such as the clays, ultrasound is used to cause dispersion. Ultrasonic baths operate on the transmission of the high frequency sound waves through the sediment or soil suspension. Bubbles form and burst creating a "sonic cavitation" that disperses the most highly aggregated particles without damage to the individual grains. Ultrasound can be used without chemical dispersants. Simple suspensions of distilled water are routinely used. Sediments or soils rich in calcium (pedocals), iron oxides (pedalfers), clays, and organic matter all disperse well in ultrasound baths. However, Mikhail and Briner (1978) insist that a step-wise pretreatment of a sample with H_2O_2, weak acid wash and sodium hypochlorite saturation before ultrasonic treatment produces the best results.

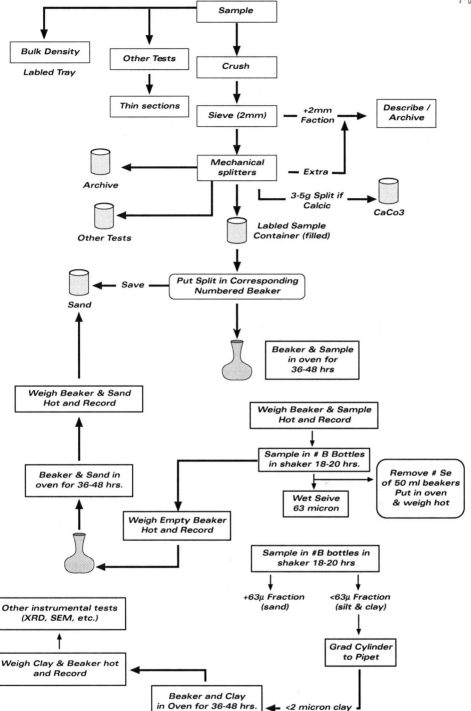

Fig. 5.2. Flowchart for standard soil and sediment processing

9.28

Mechanical sieving is the first step in PSA. Screen sieve sizes have long been standardized and the problems and limitations inherent to this method well understood (Day 1965). Reliable sieving requires the use of mechanical shakers that are sets-and-forgets such that consistent results can be expected. Pansu et al. (2001) provide a comprehensive review of grinding and sieving equipment in their text on soil analysis. If one does not use ultrasonic dispersion, sedimentation techniques follow most sieving using pipette or hydrometer methods. The settling of small particle grains such as the clay minerals follows Stoke's law in the form:

$$v = g(\rho_s - \rho_i)X^2/18\eta$$

where v (mm/s) is the settling velocity of the particle; ρ_s (kg/m^3), the particle density; ρ_i (kg/m^3) the liquid density; X (mm) the particle size; η (kg s^{-1} m^{-1}) the viscosity of the liquid and g (9.8 m/s^2) is the gravity. Stokes assumed particles to be spherical, an assumption that does not affect the results too drastically given that many particles do not meet this criteria, notably the clays. The diameter of the particle and, hence its size, is obviously more critical than sphericity. To do PSA on the sediment fraction two methods or apparatus are generally used: (1) a pipette method or (2) a hydrometer method. Both have variants and are described in detail in most works on PSA technique. Much like one's choice of dispersal techniques, the specific method is the one that most effectively accomplishes the goals of one's analysis of a specific type of archaeological sediment. All are somewhat time consuming with typical times of 20 h for clay dispersal (Rockwell 2000). Examples of each sedimentation method are presented below.

5.3
Hydrometer Method

This technique is commonly utilized to assess the clay fraction of sediment samples. The technique allows for rapid, multiple measurements from the same solution/suspension. The standard hydrometer used in the USA is the ASTM 152H model. The upper stem of the hydrometer is graduated to determine the settling depth, h, for particles of diameter, X. The settling depth (mm), h, is directly related to the graduated reading, R, (ml) on the hydrometer stem. For an ASTM 152H the empirical calculation is

$$h = -0.164R + 16.3$$

To calculate X, the variant of Stoke's law used is

$$X = \theta t 1/2$$

where t is temperature (°C), and $\theta = 1000(Bh)1/2$ where $B = 30\eta(\rho_s - \rho_i)$; the units are discussed above.

One can continue to obtain the concentration of the clay fraction in g/l from the simple calculation $C = R - R_L$, where R_L is the hydrometer reading of a blank solution. These readings are taken at the same time intervals (e.g., 1.5 h, 24 h, etc.). The term, summation percentage P, is equal to $C/C_o \times 100$ where C_o is the oven-dry weight of total sediment sample. For instance, the percentage of the clay fraction in the sample is given by:

$$P_{clay} = m \ln(2/X_T) + P_T$$

where T is the time interval and m is the slope of a summation percentage curve between two time intervals T_1, T_2. The hydrometer may be used to determine the sand fraction percentage which, plus the clay percentage, may be subtracted from 100 to obtain the silt percentage in the sample; % silt = 100 − (% clay + % sand).

5.4
The Pipette Method

The method most commonly used to determine silt and clay size fractions in a sediment sample is the sedimentation pipette method as described by Day (1965) and Green (1981). The sample sediment suspension is placed in a 1-l cylinder after 10 ml of HMP has been added together with distilled water to make up the 1-l volume. Using a Lowy pipette, a 10-ml sample is withdrawn from the cylinder at a depth of 200 mm at set time intervals. A typical sequence, suggested by Catt and Weir (1976) is as follows:

57 s	= 0.063 mm or 4 φ fraction,
3 min 48 s	= 0.031 mm or 5 φ fraction,
15 min 12 s	= 0.016 mm or 6 φ fraction,
1 h 48 s	= 0.008 mm or 7 φ fraction,
4 h 3 min 12 s	= 0.004 mm or 8 φ fraction

These aliquots are drained into pre-weighed aluminum evaporating dishes. These weighing dishes are placed in an oven at 105 °C to dry. The dishes are then weighed and the weight percentage per fraction calculated. The silt size fractions are determined similarly using established settling times from tables. Pearson et al. (1984), in a successful study of inundated archaeological sediments, used the following particle size analysis on an average 150 g sample, following Folk (1968), which was analyzed in a two-step procedure – coarse fraction then fine fraction:

Course Fraction (Steps 1–7)
1. Coarse fraction samples soaked in water for 24 h to disperse clays then dried (air).
2. Weigh sample.
3. Sample soaked, in water, to further disaggregate clays.

4. Sample wash through a 4 φ (0.067 mm) wet sieve screen to remove all the fine (silt and clay) fraction.
5. Sample dried and weighed.
6. Subtraction of this weight (5) from the prewash sample weight yields the fine fraction weight.
7. Depending upon the particular size classes of interest, e.g., five to eight 1 φ divisions, further separation of the coarse fraction is done by sieving.

Fine Fraction (Steps 8–11)

8. The fine fraction sample is dispersed with Calgon at a concentration of 2.1 g/l.
9. The sample is placed in a 1000-ml beaker and vigorously stirred.
10. Pipette samples, with a standard (Lowy) pipette, are drawn at various times (up to 8 h) at various withdrawal depths for a total of seven withdrawals/samples.
11. Each withdrawal is dried, weighed and analyzed at selected size division (e.g., 4–100 φ)

5.5
The Modified Pipette or "Fleaker" Method

A new procedure, reported by Indorante et al. (1990), has the advantage over the standard pipette method described by Day (1965) and Green (1981). The principal difference is the use of a single vessel called a "Fleaker", to disperse and pipette the sample. The fleaker is a 300-ml vessel that is a combination of the beaker and Erlenmeyer flask forms. Aliquots are taken from the fleakers in a manner similar to that described above. These are dried and weighed to arrive at estimates of the silt and clay factions in the following manner.

Five grams of soil are placed in the 300-ml fleakers that are tared to 0.001 g. Five milliliters of sodium metaphosphate solution (50 g/l) are added to the fleaker with a pipette. The mixture is then diluted to the 100-ml level, the fleaker stoppered and put on an Eberbach shaker for 8–12 h at slow speed. After shaking, the mixture is diluted further to 250 ml, stoppered again and vigorously shaken by hand for 1 min. The sample is then allowed to settle for a 1.5–2-min interval. A 20–25-ml aliquot is withdrawn at the 5 cm depth with the time noted. The mixture is, thus, repeatedly sampled in this manner with the respective samples being drained into weighing cups, typically of lightweight aluminum that are tared to 0.001 g and placed in a drying oven. The remainder of the liquid remaining in the fleaker, after the respective size samples have been withdrawn, is rinsed into a 63-μm sieve. Any sand recovered in the sieve is rinsed with de-ionized water and placed in a container to be dried overnight as the other samples. The containers are then weighed to 0.0001 g and the fractions compared to the particular particle size scheme used in the procedure (USDA, etc.; Fig. 5.1).

5.5.1
Statistical Parameters and Particle Size Analysis

Particle size analysis and the other four techniques presented in this chapter require standard parametric statistical estimators to be applied to the data they produce. These estimators include mean, variance, and standard deviation, together with parameters such as skew, sorting, etc. Some of these measures will be discussed in some depth in Chapter 8.

5.6
Point Count Analysis

MICROSCOPIC EXAMINATION

Point count analysis (PCA) is a time-honored technique developed in sedimentary petrography for the determination of the proportions of various minerals in a thin section of a rock specimen (Carver 1971). The assumption is that the results of a point count will accurately reflect the percentage of particular components – feldspar, quartz, pyroxene, pyrite, etc. within the sample and to a larger degree the percentage of that mineral (or particle size) within the source from which the sample was taken (the Delesse Relation). As implied, PCA was developed by petrographers to assess the mineral content in whole rocks using what is a sampling technique. Shackley (1975) applied point counting to archaeological sediments. In PCA studies much discussion is given to the number of grains that should be counted. This is to be expected as PCA is a sampling, n, of a larger population, N, wherein common statistical estimators such as average, mean, standard deviation, and variance are used to characterize the relationship of the sample to the population. Chapter 8 will discuss statistical procedures commonly used in archaeological geology rather than embed that discussion in this topic. Without diverging too far from this goal, it is appropriate to discuss the relationship between particle size and that of the sample size to be counted, n. It is almost a mantra in statistics, i.e., as one increases the sample size, n, the approximation of estimators for N, the population, is more assured by the Law of Large Numbers (cf. Chap. 8).

One cautionary side of this is that the result one obtains is uninformative as it represents estimator values that are so "average" as to be of no real value. If one counts enough of anything – mineral grains, pottery sherds, people, etc. – the statistical estimates of those large samples will be so homogeneous as to be unreflective of important subpopulation differences such as the difference in igneous rock types – basalt vs. rhyolite; phases of pottery; and different cultural groups. This is a conundrum present in all sampling studies – what value of n is large enough to detect important population differences, yet not so large as to mask the very differences we seek to detect and reliably estimate? Another way to express this problem is to cast it in terms of randomness vs. nonrandomness. Too large a sample, n, allows for the suppression of nonrandom trends or differences so that the results are reflective of random processes at play in all populations. About PCA, much ink has been spilled over the appropriate procedures to be used in arriving at reliable estimates of the constituents of rock or sediment aggregates as counted from a thin section or facsimile thereof. These discussions extend down to the grid type, size and microscope stage to be used. The aim is the same – count enough "points" on the slide to estimate the target population correctly.

What is the sample size, n, that best estimates grains smaller than a cobble and larger than a clay? Assuming that most samples are either rock or sediment specimens with grains of sand-size or less, then the grain count will fall somewhere between 100 and 1000. Sample counts of between 200 to 400 grains are fairly standard in studies of foraminifera, pollen and sedimentary rocks (Bryant and Holloway 1983; Krumbein and Pettijohn 1938). Krumbein and Pettijohn (1938) graphically illustrated the estimation of probable error in varying sample sizes used in PCA (Fig. 5.3). According to that analysis, it can be seen that the probable error stabilizes between 5 and 15% at about 300 grains. It is easy to see why this sample count is so commonly used in pollen studies.

5.6.1
Point Count Analysis – Sample Preparation and Microscopes

In microbotanical (pollen, phytoliths) and microfaunal (forams) studies, the sample is dispersed in solution across a counting slide which can be divided into a grid or can be covered by a glass coverslip, appropriately divided for counting purposes. Such standardized items exist for these particular analytical procedures. In geology the most commonly point-counted sample is the rock thin section. More often today the sediments examined by PCA are in their section mounts as well (Goldberg 1999). Still, it is not uncommon to see sediments dispersed as evenly as possible onto the gridded card or glass slide covered with a mounting spray cement (Gagliano et al. 1982).

Most PCA is done using stereo microscopes of relatively low power (7–30×) magnification. Depending upon the specific goals of the PCA, the microscope can be either petrographic or not. If mineralogical identification is important then a petrographic capability is required. This would be the case in studies of archaeological stone such as marbles, granites or other building or artifactual rock. It would be useful in the identification and characterization of temper found in earthenware. For instance, in thick (100 μm) thin sections, one should readily differentiate the quartz and feldspars on the basis of bright second-order interference colors under crossed polars (XPL). At standard thin section thickness, 30 μm, polarizing filters allow the full gamut of mineralogical identification (cf. Chap. 6). When examining slides for pollen and the smaller phytoliths, polarized microscopy is not required and in most cases would impede the PCA. This is because the transparency of the

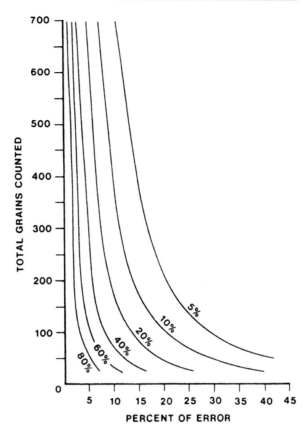

Fig. 5.3. Probable error curves for pollen count data

grains is important in making a reliable identification based on particle shape and texture (I. Rovner, unpubl.). A raster-type microscope stage (X–Y direction) is used to move the appropriate grid distance across the slide, counting grains at the grid intervals, until the selected grain sample count (n) is reached.

As Van der Plas and Tobi (1965) pointed out in their discussion, the reliability of point counting results is determined not only by the number of grains counted, but on the distance between points as well. Their point is that there is an obvious relationship between a sample's grain size and the point distance chosen for its modal analysis. For instance, it would be patently foolish to choose a point counting distance of 0.2 mm if all or most of the grains were 2 mm in diameter. The grain/point sizes will vary and therefore, the point counting protocol must be flexible to match analytical requirements of the specific case. Van der Plas and Tobi (1965) stress that "the point distance chosen should be larger than the largest grain fraction that is included in the analysis.

5.7
Organic Content Determination Methods

Other important parameters, TOC (total organic carbon), LOI (loss-on-ignition) along with mineral oxides (FeO, MgO, etc.) together with water content, are provided in most analytical packages at fee-service laboratories. Of these the organic content can be determined in most laboratories without too much difficulty. The mineral oxides are generally done with analytical methods (XRF, etc.) that may or may not be readily available in most labs. A discussion of these analytical methods is presented in Chapter 7.

5.7.1
Total Organic Carbon

Total organic carbon is the sum of organic carbon in the soil. While new instrumental methods have been developed in recent years, most soil testing laboratories routinely use one or the other of the following: (1) wet digestion; (2) colorimetric (Soil and Plant Analysis Council 2000).

The wet-oxidation digestion method uses potassium dichromate ($K_2Cr_2O_7$) with external heat (hot plate) and back titration to measure the unreacted dichromate which serves as the proxy for the carbon content. Typically 0.1–0.5 g. of soil is added to a 500-ml Erlenmeyer along with 15 ml of dichromate reagent. The mixture is then boiled for 30 min. After cooling, 3 drops of the indicator solution n-phenylanthranilic acid is added and the mixture titrated against Mohr's salt solution, ferrous ammonium sulfate, until a color change from violet to bright green is observed. The % organic matter = {[meq $K_2Cr_2O_7$ – meq $FeSO_4$ × 0.3]/g soil} × 1.15. Here, meq is milliequivalents.

The colorimetric procedure uses the property of absorbance as discussed in detail below in reference to phosphorus determination. Here, 2.0 g of soil is placed in a test tube to which 20 ml of sodium dichromate solution is added. The reaction is exothermic and after the mixture has cooled (40 min–1 h), it is allowed to stand for 8 h. Aliquots of the clarified solution are then transferred to the spectrometer for absorbance measurements at 645 nm wavelength (Soil and Plant Analysis Council 2000, 178). The values obtained, when plotted in comparison to standard curves yield the organic content.

5.7.2
Loss-On-Ignition

This method has become comparable in reporting to that of the methods just discussed. It is equally effective for the analysis of soils containing low to very high organic matter (OM; Soil and Plant Analysis Council 2000). It has a sensitivity of 0.2–0.5% organic matter. A sample of 5–10 g soil weight (W) is placed in an ashing vessel. The vessel is first placed in a drying oven at 105 °C for 4 h. Once cooled, the sample is weighed to the nearest

0.01 g and then placed in a muffle furnace at 400 °C for 4 h. The sample is removed, cooled and weighed again. To calculate the % organic matter one uses the formula:

$$\%OM = [(W_{105} - W_{400}) \times 100]/W_{105}$$

5.8
Bulk Density

Bulk density is the average dry weight per unit volume of soil. Depending on the nature of the sample, fine or coarse-grained, the principal method is to determine the matrix bulk density by addition of paraffin and subsequent measurements. The sample is first dried and weighed, then soaked in paraffin, dried and weighed again. This determines the weight (and volume) of the paraffin. Finally, the sample is immersed in water and weighed. Bulk density, in g/cm³, is calculated thus:

$$\text{Bulk density} = (\rho_w)(WC_d)/WC_p - WC_{pw} + [(W_p)(\rho_w)/\rho_p]$$

where ρ_w is the density of water; WC_d is the dry weight in air; WC_p is the waxed weight in air; $WCpw$ is the waxed weight in water; Wp is the weight of the paraffin; ρ_p is the density of the paraffin.

For fine-grained samples (size $\ll$ 2 mm) the procedure is straightforward. With coarser, gravel-rich samples, some error can be introduced by the size of the large clasts. A sample should be broken apart to determine the amount of these clasts within a sample. As a result, the bulk density so determined is more an average value and less a specific value for an individual sample. Elevated bulk density values are typical of older soils with significant clay and carbonate volumes (Soil and Plant Analysis Council 2000).

5.9
Carbonate Content

The amount of calcium carbonate in aridisols and lacustrine sediments can be significant and must be accurately measured. Like the straightforward manner of basic mineral determination of the introductory geology laboratory, the use of hydrochloric acid is required. In this more rigorous determination the HCl release of CO_2 gas is collected and measured by the use of a sealed flask and graduated column apparatus called the Chittick method (Singer and Janitsky 1986). The evolved gas displaces a fluid (colored water) into the graduated column. This value is divided by the original sample weight to determine the carbonate content which is generally reported as percent (Singer and Janitsky 1986).

5.10
Phosphorus Analysis

As Joseph Schuldenrein (1995) states "P (phosphorus) appears to be the most generic indicator of anthropogenic input . . ." into archaeological sediments. The importance of phosphorus analysis in archaeology was first recognized by Arrhenius (1931), but it was R.C. Eidt who established the systematic use of the phosphate method in evaluating anthropogenic sediments (Schuldenrein 1995). Eidt, his coworkers and students, established the processes chemistry and chronology of phosphorus as phosphate compounds in archaeological sediments (Eidt 1973, 1977, 1985; Sjöberg 1976; Woods 1977). The assumption is that the magnitude of human impact on a landscape can be demonstrated by the total phosphorus concentration so that archaeological sites and inter-site activity areas/features can be isolated. Phosphorus occurs in sediments as either bound (nonlabile) or unbound/free (labile) phosphate compounds. The labile forms correspond to Al and Fe phosphates. Bound or occluded forms of Al and Fe phosphates develop as inorganic compounds in soil horizons.

Inorganic compounds are (1) those containing calcium and (2) those containing iron and aluminum. The solubility of inorganic calcium compounds increases with the increased simplicity of the compound, but the simpler compounds are only available in very small quantities and they easily revert to insoluble states. The fixation of phosphates, when the pH value is in the range of a little less than 8.0 and 8.5, is mostly in the form of calcium phosphates. In alkaline soils phosphates will react with the calcium ion and with calcium carbonate. The next step includes a conversion of the calcium phosphate ($Ca_3(PO_4)_2$) to insoluble compounds, such as hydroxy-, oxy-, carbonate-, and fluor-apatite (insoluble apatite compounds are formed at pH levels of above 7.0) compounds, the latter being the most insoluble compound known. An excessive amount of calcium carbonate ($CaCO_3$) produces very insoluble compounds. At a pH level of between 6.0 and 7.0 phosphate fixation is at its minimum.

It is not within the scope of this volume to fully describe the extremely complex and incompletely investigated phosphorus cycle in sediments. It is important to mention the potential sources of both anthropogenic and nonanthropogenic sources such as geology and soils. After mineral sources in sediments and soils, the bulk of phosphorus is from organic matter such as dead plants and animals that decompose in the soil. Phosphorus is an important energy source in organisms, particularly as a component of the ATP (adenosine triphosphate) molecule. As a result, all plants and animals contain significant amounts of phosphorus in their tissues which is incorporated into sediments after their death. In addition, humans and animals excrete large amounts of phosphorus as both feces and urine. Plant and animal decay, burials, refuse coupled with urine and excrement increase the

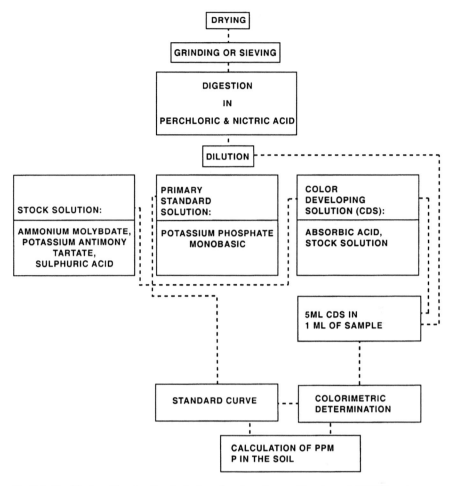

Fig. 5.4. Perchloric acid extraction technique for phosphorus. (Courtesy of Alf Sjöberg)

phosphorus content of soils above ambient levels. The accumulation of this phosphorus in anthropogenic soils, if it can be fixed into insoluble forms, can have long residence times – millennia in many cases – and provide a relatively reliable geochemical marker of archaeological interest.

5.11
Determination of Total Phosphorus by Perchloric Digestion

Information on this method is found in Eidt and Woods (1974). The methodological procedure in the chemical analysis is described through the steps noted in Fig. 5.4. The method extracts total phosphorus, i.e., both organic and inorganic phosphorus compounds. The accuracy level of the method of the analysis ranges between 95–98% recovery. All samples are dried overnight in an oven at a temperature of around 50 °C. Samples consisting of very fine particle sizes are ground directly after having been dried, while samples with coarser particle sizes have to be screened. The choice of sieve size was made so that the sieved sample would have the same particle size as the ground one. A half gram of each sample is put into a graduated digestion tube to which is added 2 ml of perchloric acid ($HClO_4$) and nitric acid (HNO_3) respectively.

The tubes are then placed in an aluminum heating rack under a perchloric acid fume hood (Blanchar et al. 1965). The HNO_3 is boiled off at a temperature of 170 °C. After a stepwise increase to 225 °C the remaining $HClO_4$ is left to boil for 1 h. The tubes are checked regularly to prevent them boiling dry. After the digestion, the volume of each tube is raised to 50 ml with distilled H_2O, stirred, and left to rest for a period of at least 12 h. A stock solution is made consisting of (1) 9.0 g ammonium molybdate [$(NH_4)_6Mo_7O_{24}$ $4H_2O$] and 0.2 g potassium antimony tartrate [$K(SbO) C_4H_4O_6/2H_2O$], which are mixed in 500 ml of H_2O, and (2) 65 ml sulfuric acid (H_2SO_4) and 335 ml of H_2O. Initially mixed separately, the (1) and (2) reagents are mixed together and the volume raised to 1000 ml with H_2O.

A primary standard solution with a concentration of 1000 μg/g P is made by dissolving 4.387 g potassium phosphate monobasic (KH_2PO_4) in sufficient H_2O to yield 1000 ml. A secondary standard solution, with a concentration of 10 μg/g P, is made by adding 40 ml $HClO_4$ to 10 ml of the primary standard solution and dilution with H_2O to a volume of 1000 ml. Working standard solutions ranging from 1 to 6 μg/g P are made from the secondary standard solution. The color-developing solution, freshly made for every batch of 100 samples, consists of 2.0 g ascorbic acid mixed with 200 ml of the stock solution and raised to 1000 ml with H_2O. Five milliliters of the color-developing solution is added to 1 ml of each of the working standard solutions as well as to 1 ml of the samples. After 1 h their absorbencies are read on a colorimeter at a wavelength of 740 nm. The absorbance of the standards is plotted against the known concentration. From this graph the concentrations of the soil samples were calculated (Fig. 5.4).

The chemical analysis as just described is accurate and has good reproducibility. A linear regression of P content and absorbance gives a correlation coefficient (r) of 0.9998 (cf. Chap. 8 for a detailed discussion of linear regression). The regression line plotted in Fig. 5.5 is a linear relationship between absorbance and concentration judging from the r value of 0.9998 and the straight line. Beer's law states the following: "the fraction of light absorbed is directly proportional to the concentration of the colored constituent. The phosphate determination conforms closely to Beer's law: the slope of the regression line is just below optimal linearity as is expressed by the regression equation $y = 0.0957x + 0.0028$. A direct proportionality, as stated by Beer's law, would mean $y = 0.1x$.

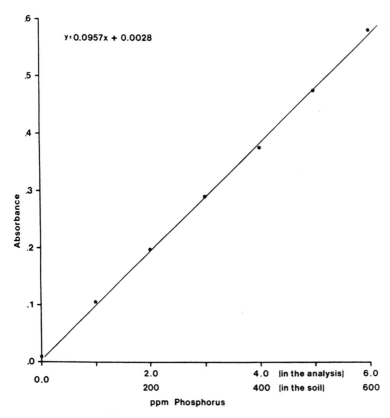

Fig. 5.5. Calibration curve – regression – for phosphorus vs. optical absorbance. (Courtesy of Alf Sjöberg)

5.12
Absolute Phosphate Analysis Versus Qualitative Color Tests

Total phosphorus-bound fraction analysis vs. those of labile or less bound phosphates has little or no comparative value. The latter phosphates can be qualitatively detected by methods similar to the familiar litmus strip test for acidic/basic solutions. In the qualitative phosphate test one measures the intensity of color produced in a spot test on paper or in a small tube. While somewhat subjective and decidedly not quantitative, the color test has economy and ease of application in a field setting on its side. At the site of Great Zimbabwe, researchers from Lund University used both quantitative, bound phosphate analysis as well as the field color tests with good effect (Paul J.J. Sinclair, pers. comm.; Sinclair, 1991, Sinclair and Pétren, unpubl.). The Swedish team, in extensive coring tests, extracted phosphates in column samples to measure the extent, vertical and horizontal, of the famous African Iron Age site (Fig. 5.6). Intercomparison of their data – relative color

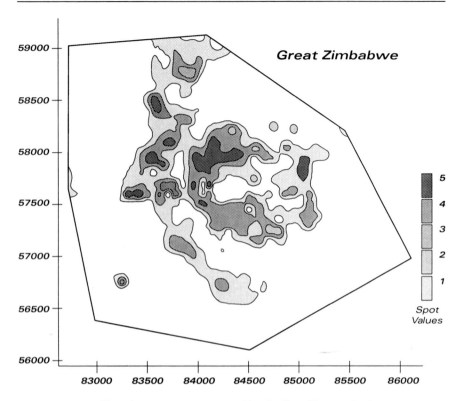

Phosphate content measured by the Spot Test method,
with cored areas and perimiter walls.

Fig. 5.6. Great Zimbabwe phosphate distribution map (Adapted from Sinclair and Pétren, unpubl.)

intensity and the absolute colorimetric determination – showed similar findings in the measurement of site extent.

Cavanagh et al. (1988) utilized a "cheap, robust" colorimetric phosphate procedure developed by Craddock et al. (1985). To assess site parameters at Laconia, Greece, Craddock et al. (1985) developed the molybdenum-blue method to analyze multiple samples in a relatively short time. Like the simpler, qualitative tests used at Great Zimbabwe, the procedure used at Laconia examined the total phosphorus present without the extraction of the bound (P_2O) "total" component as accomplished by the perchloric acid technique described. The molybdenum-blue method is described as follows: (1) beginning with excavated/cored 10-g soil/sediment samples that are (2) air dried and screened through 1-mm metal sieves, (3) 1 g subsamples are digested in (4) 5 ml of 2 N HCl boiled for 10 min, (5) a 0.2-ml aliquot is removed and 10 ml of molybdenum-blue reagent is added and (6) the liquid is analyzed with a colorimeter standardized to zero optical density with a phosphorus-free reagent. Like the perchloric approach, a set of eight or more reference samples of known concentrations are used to produce a calibration regression line. If percent transmittance (T) is measured, this can be converted to optical density by the formula:

Optical density $= -\log 10^{T}$

The molybdenum blue reagent is mixed from 65 ml H_2SO_4 (6 N), 37.5 ml ammonium molybdate (40 g/l), 12.5 ml potassium actimonyl tartrate (2.743 g/l) and 1.32 ascorbic acid (solid) as suggested by Murphy and Riley (1962).

5.13
Colorimetry and Spectrophotometry

For the geoarchaeologist, relatively "quick and dirty" methods exist for ascertaining total phosphorus without recourse to elaborate laboratory techniques such as perchloric digestion. This latter technique is essential for a thorough, quantitative determination of the bound and "condensed" phosphate fractions. One other component of the total phosphorus is that of the labile or orthophosphate fraction. This is helpful to archaeologists as it can be readily determined in the field.

The most common procedure is acid hydrolysis of the organic and condensed inorganic forms (meta-, pyro-and other polyphosphates) to reactive orthophosphate (Hach Water Analysis Handbook 1992). Pretreatment with an acid (sulfuric, nitric) and heat provides the conditions for hydrolysis of the inorganic fraction: (a) organic phosphates are converted to orthophosphate by heating with acid (acid oxidation) and persulfate (b).

(a) $Na_4P_2O_7 + 2H_2SO_4 + H_2O \rightarrow 2H_3PO_4 + 2Na + 2SO_4^{2-}$

(b) $R-O-P-R' + K_2S_2O_8 + H_2SO_4 \rightarrow H_3PO_4 + 2K^+ + 3SO_4^{2-}$

In (b) R and R' represent various organic groups. If one determines the total phosphorus by the second method (b), it can then be followed by the reactive phosphorus test. By subtracting that amount from the total phosphorus result one can determine the organically bound phosphate which is a good indicator of cultural deposits of archaeological interest. The analysis of reactive phosphorus – the orthophosphate – can be done using a field spectrophotometer. Orthophosphate absorbs at the 890 nm wavelength in the Hach field spectrophotometer and produces reliable measurements in mg/l of PO_4^{3-}. Using the same device, one can assess the total phosphorus (PO_4) using either the acid persulfate digestion method or the ascorbic acid method (Hach Water Analysis Handbook 1992).

Colorimetry and spectrophotometry rely on straightforward principles where the color of a molecule in solution, such as phosphorus, depends on the wavelength of light it absorbs. When a sample solution is exposed to white light, certain wavelengths are absorbed and the remaining are transmitted to the eye or a photometer. In our discussion of the determination of total phosphorus by perchloric acid digestion, the final measurement is to test absorbance in a light cell where the absorbance and concentration of the unknown is determined from an absorbance vs. concentration curve, for the element of interest, in our case P, constructed by use of standards. The absorbance A, is generally defined as the log (P_o/P) where P_o = incident light intensity and P = intensity after absorption. Using Beer's law, $A = a \cdot b \cdot c$, where a = molar absorptivity (%), b = path length (μm) and c = concentration (g/ml) . We hold a, b constant and measure c, concentration by the spectrometer.

5.14
Palynology – Study of Archaeological Pollen

Fossil pollen grains and grass spores are preserved in many soils and sediments that are not overly acidic. These elementary reproductive particles are produced in large amounts by both trees' shrubs, and grasses. Some trees like pines produce millions of pollen grains annually. Other plants, by contrast, are more meager in their output, such as clover (200 grains per anther). The grains themselves are larger than clay particles (10–100+μm) and are amenable to optical microscopic study. Pollen grains are tough customers being preserved by a very resistant exine* made of sporopollenin. The exine defines the basic morphology of the pollen grain and is distinctive in most higher plants. The palynologist inspects the grains for shape, size, surface texture and apertures. As a result of over a hundred years of taxonomic studies of pollen, beginning with Lagerheim and his student von Post (Bryant 1989), in the early 1900s, we enjoy a broad and reliable repertoire of pollen identification keys (Faegri and Iversen 1975; Bryant 1989).

Von Post set forth five principles for pollen studies which remain valid today:

1. Many plants produce pollen or spores in great quantities, which are dispersed by wind currents.
2. Pollen and spores have very durable outer walls (exine) that can survive long periods of time.
3. The morphological features of pollen and spores remain consistent within a species; different species product their own unique forms.
4. Each pollen- or spore-producing plant is restricted in its distribution by ecological conditions such as moisture, temperature, and soil type.
5. Most wind-blown pollen falls to the earth's surface within a small radius, roughly 50–100 km from where it is dispersed.

As von Post observed, pollen is often produced in significant quantities that (1) survive in sediments for long periods; (2) have distinctive, often species-

*EXINE – OUTER COAT OF A SPORE

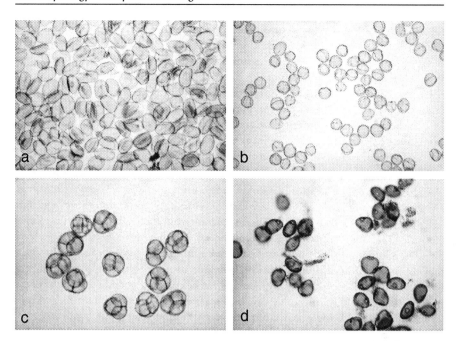

Fig. 5.7. Four pollen grain varieties: a bean, b triporate (oak), c tetrad (rhododendron, d triporate (beech). (Courtesy of Prof. George Brook)

specific forms and (3) are distributed over significant areas, mostly by the wind as "pollen rain". The pollen of trees is termed arboreal or AP; that of shrubs and herbs, together with some grasses, nonarboreal (NAP). Some common types of pollen are shown in Fig. 5.7. Bryant (1978) lists three factors that influence the preservation of pollen grains in sediments and soils: (1) mechanical, (2) chemical, and (3) biological. Mechanical damage occurs during dispersal and deposition generally by crushing or abrasion of the grain. Chemical corrosion of the exine is related to the pH of the depositional environment such that sediments with pH 6.0 or higher degrade pollen – alkaline soils that fix phosphates degrade pollen. Fungi and bacteria can breakdown the pollen. Susceptibility of pollen to corrosion results in the differential preservation of various species. Pine, although produced in large amounts, along with oak, elm, and birch, can be over represented in pollen spectra by a factor of 4, whereas alder and beech are direct reflections, percentage-wise, of their occurrence in the forest (Andersen 1986).

Archaeologists and scientists interested in past environments and must recognize the various factors that obscure recovered pollen percentages as representational of vegetation composition. Key factors are the biological productivity of the respective plants, dispersal modes and patterns, and grain survivability. For the purposes of archaeological geology, it is more important to examine methods for the recovery and extraction of pollen grains, than to reprise the specialized discussions of varieties of pollen, their identification, and the various methodological approaches to reconciling pollen spectra to vegetational reality in the past.

Pollen can and should be considered as a sediment. It is an organic component of the sedimentary matrix and *sensu stricto* a defining element of "soil". In paleosols, pollen is preserved subject to soil geochemistry. Within the peds of a soil, the pollen grains are found with the silt-sized particles. Such pollen grains are not recoverable with the methods typically used to separate the coarse sediment fraction. Pollen grains are separated by dispersal and digestion procedures. Unlike the techniques used in the separation and preservation of the clay, those used for pollen recovery utilize more caustic reagents such as potassium hydroxide, hydrochloric acid and the high reactive hydrofluoric acid. Acquiring sediments for pollen extraction and analysis can be done in conjunction with the study and description of a soil/sediment profile. Similar to our example of obtaining a soil monolith, a block of material is removed by use of a trowel or small shovel. Such a block or vertical sequence of blocks can be returned to the laboratory for subsequent, systematic removal of subsamples every few centimeters as a general rule.

Coring is a common means of retrieving continuous sediment columns. Of the various coring procedures described in Chapter 4, the one most appropriate for the specific study depends in great part on the sediment/soil being sampled. Cores are taken in wetlands such as bogs, sinks and shallow lakes where sediments are conducive to pollen preservation. Short, open barrel corers can simply be pushed into soft sediments. Deep cores require mechanical or hydraulic assistance. In drier sediments the rotary, percussion or drive corers are useful. After recovery the sediment column is subsampled at selected intervals (ex. 10 cm).

If the sediment is in the form of a core then some care must be taken to note not only the sampling interval, but where that interval is in relation to the *top* of the core. Most investigators stress that the sample be labeled, in terms of distance (cm, mm, etc.), from the top of the core. Until just recently, the author was in complete concurrence with this procedure. Having used geological coring as a sediment sampling method technique for some time, it now seems that using the top of core, as the point of reference, may not be as appropriate as previously thought. In situations of single cores using the top of the core as "zero" may have merit. In the case of multiple cores across an uneven topography, the stratigraphic relationship between cores, in an absolute sense, may be difficult to determine without some measure of surface elevation, so that each core can be placed in correct relationship to its neighbor. One core must be a reference elevation datum and the others placed in relation to it.

Where such elevation data are lacking it is suggested here that the measurement of "zero" in the core may be more useful if taken from the *bottom* or tip of the core. There is nothing arbitrary about the bottom point of a core. In sampling wetland or submerged sediments, it is not uncommon to lose portions of upper sediments due to suspension/dispersion of light fractions or simple unconsolidation of the sediment by the coring device. In these cases one does not have a true surface measurement that is coincident with the observed top of the sediment in the core sample. By measuring from the core tip some consistency is assured. In the case of varying surface topography, this procedure "normalizes" the disparity in strata by removing the offset in altitude so that these strata will be observed at their same relative positions in the core sample, regardless of surface elevation.

Additional caution should be taken in the removal of sediment, whether from a core or profile. Scrape the sampling site's surface removing the surface material with a spatula (trowel, etc.) that is *cleaned* between samples. Take uniform samples of 0.5–1.0 cm³ (a few grams) in size. A typical extraction procedure is shown in Table 5.2. The sequence of various reagents and dispersals removes the sediment matrix containing the pollen and spores. After the pollen grains and spores are extracted, they are stained and mounted. A typical sample area, such as 22 × 22 mm is called a "strew" (Rich 1999).

As with point counting, the pollen grain count results in a relative sample frequency. This sample pollen frequency must be converted to absolute pollen frequency to compensate for disparities in pollen influx at different times such as the difference between fluences of 1000 and 100,000 cm⁻², although both represent 90 percentile samples. Davis (1969) suggested a novel method to do this by amending or "spiking" the samples with a tracer pollen type in known amounts. Using this method the number of pollen grains of each pollen type can be ascertained relative to the known ratio of the tracer species.

The spike procedure may use something like exotic pollen grains like those of *Eucalyptus*. The concentration of any pollen taxon can then be measured in ratio-proportion to this "exotic" spike by the simple relation:

Table 5.2. Sample preparation for pollen analysis. (Courtesy of George Brook, University of Georgia, Geography Department)

(i) Boil sediment with 10% NaOH or KOH to remove soluble humic acids and to break the sediment down
(ii) Sieve to remove coarse debris
(iii) Treat with cold 1.0% HCl to remove carbonates
(iv) Treat with either hot 48% HF for a short time (up to 1 h or cold 48% HF for a long time (up to 24 h) to remove silicates (silt, clay, diatoms, etc.)
(v) Treat with hot acetolysis mixture (9 parts acetic anhydride: 1 part concentrated H_2SO_4) to hydrolyze cellulose
(vi) After neutralization, stain the residue with safranin or basic fuchsin
(vii) Mount the residue in a suitable medium with a refractive index (1.55 which is that of pollen grains. An excellent medium is silicone oil (R.I. = 1.4)

$$\frac{\text{Observed spike grains (a)}}{\text{Total spike grains (b)}} = \frac{\text{Observed grains of taxon X (c)}}{\text{Total particles of taxon X (d)}}$$

where one simply solves for (d) total particles of taxon X.

Of course, using a spike of *Eucalyptus* pollen is precluded in areas where the species occurs naturally or has been introduced (e.g., southern California). Shane (1992) suggests using $7-10 \times 10^4$ spike grains/ml such that by adding 0.5–1.0 ml of this spike to a 0.5 cm^3 sediment sample will yield an "ideal" 2 grains of the pollen spike to the sample (Maher 1981). In counting a spiked sample, one procedure is to use a hemocytometer, a thick, specialized slide with apertures (2) for introduction of aliquots into two, separate grids. Once the sample aliquot is placed in the hemocytometer, the grains along the parallel sides of those grid squares marked by "X" are counted and listed in pairs. A total of 30–40 grid squares are counted. The volume/square is about 0.1 mm^3 so 10 squares equal 1 ml by volume. One hopes to observe at least 300 grains in a count of at least 5 squares to assure representation statistics. If the counts done by a palynology laboratory, show a paucity of pollen grains (~300 grains/sample) they will generally advise against extensive analyses.

5.15
Phytoliths for Archaeology

Rovner (1983, 1988) describes phytoliths as botanical microfossils that can provide significant paleoenvironmental and archaeobotanical information. Phytoliths are mineral deposits (calcium, silica) that form in and between plant cells (Mulholland and Rapp 1992). Of the two mineralogical varieties of phytoliths – calcium and silica or "opal" – the latter has formed the bulk of most modern phytolith studies. While both calcium and opal phytoliths exist as either amorphorus forms of cell-morphology (cystoliths"), the opal phytoliths exist longer in sedimentary contexts. Fossil opal phytoliths have been reported in sedimentary rock millions of year old (Jones 1964). Bryant (1974) has reported both calcium and opal phytoliths in ancient coprolytes, but the former are not frequent in sediments. Bukry (1979) has reported opal phytoliths preserved in deep ocean sediments.

In 1836, Christian Gollried Ehrenberg began the systematic study of phytoliths or phytolitharia which he published in 1845, 10 years after the first published work on phytoliths by another German scientist named Struve (1835). His subsequent, extensive collection and study of phytoliths from Europe, Africa, Asia and America led to Ehrenberg's *Mikrogeologie* (1854). Ehrenberg's lead inspired other German researchers to continue phytolith studies in sediments and expanded observations of phytolith sources (Grob 1896). From the beginning the link between phytolith studies and that of paleoenvironments, a key interest of modern archaeology, was articulated by Ehrenberg and those who came after him. Although increasingly more common, phytolith studies of archaeological sediments have yet to reach the level of acceptance of palynological studies. This situation is rapidly changing and the use of phytolith studies has proved to be of diverse benefit to both archaeology and geology (Twiss 2001).

5.16
Phytolith Identification and Morphology

Traditionally, the identification of phytoliths has been based on typologies. The most typical shapes have been given significance for the identification of plant taxonomy. Phytolith populations are complex and stochastic – suffering from a multiplicity of shapes, subtle and not so subtle morphological variation, and duplication (redundancy) of basic forms among taxa (I. Rovner, unpubl.). In this form of systematics, differences between phytoliths are ignored in favor of their similarities. Phytoliths are grouped rather than differentiated into smaller taxonomic categories such as genus and species. Furthermore, most measurements taken are the length and width since they are the easiest to measure manually (I. Rovner, unpubl.). Opal phytoliths are produced by angiosperms, gymnosperms and pteridophytes (Piperno 1988). They are produced by both monocotyledons and dicotyledons. Plant families producing opal phytoliths include both Poaceae and Gramineae grasses; Cyperaceae (sedges); Ulmaceae (elm); Fabaceae or Leguminosae (beans); Cucurbitaceae (squashes); and Asteraceae or Compositae (sunflower) (Mulholland and Rapp 1992). Still phytolith production is not universal and even when plants do so the phytoliths are morphous and are not distinctive.

The immense variety in shape stems both from intra-plant taxonomic differences as well as simple cellular differences between different parts of the plant from a specific taxa. In addition, plants produce vast amounts of these cell casts as they die and decompose in the soil. It should be noted that two major differences exist between phytoliths and pollen besides long-term preservation: (1) phytoliths are less taxonomic-specific than pollen and (2) they are more likely to be deposited in situ rather than broadcast over large areas like many forms of pollen. When phytoliths can be reliably identified, they make up stable components of archaeological sediments (Pearsall and Trimball 1984; Pearsall 1989).

The size and shape of phytoliths vary depending on the plant in which they are deposited and the area of deposition within the plant. Monosilic acid is carried into the tissues of the plant in groundwater and is usually deposited in stems, leaves, and inflorescences. Eventually the silica is precipitated forming opaline silica ($SiO_2.nH_2O$) in the shape of the area that contains it. These areas lie within or between cells. This forms the cystalith or morphocast of the cell. Upon decay of this plant, the phytoliths become a component of the soil which can then be studied to determine the past floral assemblages of the immediate area as Ehrenberg had foreshadowed. Guntz and Grob (Grob 1896) identified and taxonomically classified more phytolith varieties than their predecessor. Guntz alone studied 130 species of grass in an attempt to link the cell morphology of leaves to the area's climate in which they grew (Powers 1989).

The grass family is a monocotyledon taxon that produces an abundance of phytoliths and is also a good indicator of climatic changes. These phytoliths have been subdivided into general long and short cell categories. Distinctive grass short cells have been further subdivided into festucoid, panicoid, and chloridoid classes (Twiss et al. 1969). Festucoids include grasses that prefer cool and moist conditions and include Festuceae, Hordeae, Aeneae, and Agrostideae. Their shapes range from trapezoid, rectangular, elliptical, and acicular to crescent, crenate, and oblong. Chloridoid grasses grow under drier and warmer conditions. Their shapes are less diverse, being a basic saddle shape. Chloridoid grasses occur in Chlorideae, Ergrosteae, and Sproboleae. Panicoids include the other grass tribes of Andropogoneae, Paniceae, Maydeae, Isachneae, and Oryzeae which are located in warm, moist environments. Their short cells contain the most distinct shapes such as bilobates, cross-bodies and crenate-shapes. The grass short cells and their morphology are illustrated and discussed in detail by Brown (1984). Twiss has pointed out that for grass phytolith classification it is the costal short cells that are the most diagnostic (2001). Unfortunately, these cells comprise only a small fraction (~5%) of the total siliceous residue (Twiss 2001). Most of the residue used in a phytolith sample is composed of the relatively nondiagnostic epidermal cells. Fragments of the epidermis of leaves, stems, and husks (inflorescence bracts – hair cells, papillae, etc.) offer today's worker improved chances for identification success.

In terms of agriculture, many European grasses fall into the festucoid category while maize contains panicoid lobate shapes. Rice contains a type of phytolith known as bulliforms while parts of other cultigens such as the rind of squash, the leaf of bean, and the sunflower as well as maize form unique shapes that were identified by Bozarth (1986, 1993). Squash for example, forms spheroid phytoliths covered with contiguous concavities. Weed species that thrive in areas of agricultural abandonment often contain chloridoid short cells. The identification of phytoliths from these species remain important to the study of archaeology. Trees and shrubs also produce phytoliths. The shapes vary greatly and need further study to identify their taxa and origin. Grasses are largely unidentifiable below family level.

5.17
Phytolith Extraction and Counting

The technique of phytolith extraction set forth by Rovner (I. Rovner, pers. comm.) was used on the soil samples collected from the site. It represents a relatively simple, but labor intensive and time consuming procedure which is necessary to separate phytoliths from the other components of the soil (Table 5.3). Approximately 30 g of soil is necessary for processing using this procedure. *FOR SEPARATION*

Once the phytoliths have been successfully extracted, one proceeds to identify and count as many recognizable forms as possible at family or subfamily i.e., taxonomic levels. Most researchers prior to the 1990s relied on light microscopy to identify and count phytoliths. The method most commonly

Table 5.3. Process of phytolith extraction. (After Owens 1997)

Step	Reason
Distilled Water	Removes organic material by flotation
Centrifuged	Allows phytoliths to settle to bottom
Decanted	Removes excess water
Repeat 3 times	
Bleach	Dissolves remaining organic material
Soaked overnight	
Distilled water	Rinses off bleach
Centrifuged	
Decanted	
Repeat rinse 3 times	
HCl	Removes CaCO$_3$
Centrifuged	
Decanted	
Distilled water	Rinses off HCl
Centrifuged	
Decanted	
Repeat rinse 3 times	
Na metaphosphate	Removes clay minerals
Centrifuged	
Decanted	
Distilled water	Rinses off Na metaphosphate
Centrifuged	
Decanted	
Repeat rinse 3 times	
Dried overnight	Removes moisture
Placed in vials	
Heavy liquid (ZnBr or Na metatungstate) sp.g.2.35	Floats off phytoliths
Centrifuged	
Decanted	
Distilled water	
Centrifuged	
Decanted	
Repeat rinse 3 times	
Placed in vials	

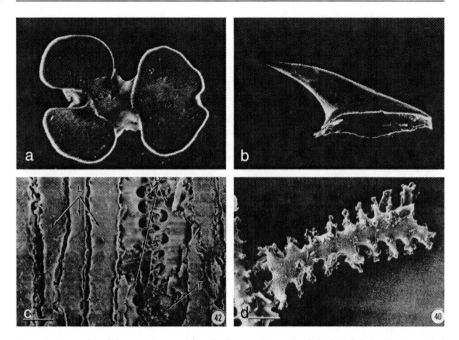

Fig. 5.8. Four phytolith varieties: **a** chlorodoid grass "cross-body" phytolith, **b** disarticulated prickle phytolith from the wheat *Triticum dicoccum*, **c** epidermal long-cell (*L*) and trichome base (*T*) phytoliths from the wheat *T. aestivum*, **d** subepidermal rod-shaped phytolith from the wheat *T. dicoccum*. (Courtesy of Prof. Irvin Rovner)

used is the "quick scan" under magnifications of 300–500×. The slide containing the phytoliths is raster-scanned while the observer tallies all phytoliths present as to known and unknown types. The use of scanning electron microscopy (SEM) and confocal technologies is changing phytolith analytical procedures, particularly with regard to the identification of phytolith morphology (Fig. 5.8). Both technologies are superior imaging systems compared to traditional optical microscopy. SEM is well established and confocal-laser (CL) microscopy has the like advantage of imaging the cell cast in great detail in storable digital, isometric images. The drawback to these higher resolution imaging techniques is scale and speed.

Because the SEM and confocal systems focus on micron–submicron scales, the investigator does not examine as many phytoliths during a scan. The identifications are more confidently made, but the overall number of individual "grains" counted is minuscule compared to the optical procedures. Russ and Rovner (1989) have suggested the use of computer-assisted image analysis to achieve reliable, less subjective phytolith identification as well as more rapid assessment of total numbers of phytoliths per sample.

5.18
Resin Impregnation and Micromorphology

Goldberg (1995) has used resin impregnation of soil/sediment blocks as a precursor to producing thin sections of the clastic materials for use in micromorphology. This procedure is finding increasing application in the geoarchaeological studies of site deposits (Courty et al. 1989). At the thin section scale, undisturbed, oriented samples are examined by the microscope – petrographic or otherwise (SEM; CL) – allowing for observation of the composition in mineralogical terms; texture and fabric – the geometric relationships among constituents (Courty et al. 1989). Microstratigraphic changes can reflect changes in depositional processes.

Undisturbed sample soil/sediment blocks are taken from the field profile. These samples are then impregnated in a vacuum chamber with epoxy resin. In this matrix, after hardening, the sample can be treated like a rock sample and sawn to produce viewable thin sections of standard dimensions. This technique was used in the example study outlined in the following section.

5.19
Applications of Techniques – An Archaeological Example

5.19.1
Late Quaternary Sediments and Soils of the Great Plains, Missouri and Illinois

Highway construction in Northwest Missouri exposed a stratified archaeological site that was found to contain a range of lithic materials extending into the Paleo-Indian Period (Reagan et al. 1978). The excavations carried out in the mid-1970s were characterized by the extensive use of geoarchaeological techniques to describe the deposits. The site, now destroyed, was called the Shriver site and was located in the northwestern prairie section of Missouri on a plain characterized by loess and glacial drift deposits overlying shale and limestone bedrock (Dort 1978). Its discovery and its distinctive lithic assemblage raised the possibility of it being a so-called Pre-Clovis site. This assignation was based on the artifactual stratification wherein a Folsum Period occupation (ca. 11,000 B.P.) was superposed on a deeper (~40 cm) assemblage of bifacial and unifacial flint artifacts containing no lanceolate projectile points (Reagan et al. 1978). Because of the unusual nature of the archaeological assemblage, the excavators conducted extensive geoarchaeological studies to include geological/geochemical, soil/sediment, pollen, and geochronological procedures. The author was involved in the geochronological aspect of the study utilizing thermoluminescence (TL) to investigate flints suspected of being heat-treated (Reagan et al. 1978).

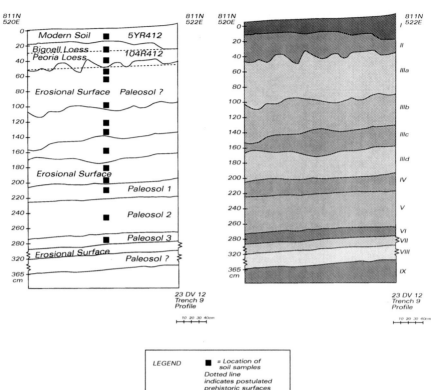

Fig. 5.9. Lithostratigraphy of the Shriver site, Missouri. Trench 9

Excavations – hand and mechanical – and later analyses by Dort and Johannsen (Dort 1978; Reagan et al. 1978), identified a total of 13 stratigraphically identifiable units (Fig. 5.9). As indicated on the profile shown in Fig. 5.9, Dort confirmed the presence of two loess units, the Bignell and Peoria Loesses. These were identified in the upper part of the stratigraphic section – units 1–II. At the base of the Peoria Loess, a unit dated by Caspall to 13,000–18,000 years, the older lithic assemblage was found. In addition, an irregular erosional surface and associated paleosol was identified. Other erosional surfaces, as well as paleosols, likely extending back to the Penultimate

Table 5.4. Soil chemical analysis of selected horizons in Trench 9, Shriver Site, northwestern Missouri

Sample no.	Depth (cm)	pH	Cation exchange capacity	LBS/A[a]				O.M. (%)
				Ca	Mg	K	P^2O^5	
107-9-1-1	10–15	5.3	22.5	5500	860	600	147	4.9
107-9-2-1	30–35	5.2	22.0	5000	970	600	41	2.7
107-9-2-2	45–50	4.7	23.0	4900	1000	600	32	2.3
107-9-3-1	55–60	4.5	25.5	5500	100	570	23	2.0
107-9-3-2	65–75	4.4	26.5	6100	1000	398	27	0.9
107-9-3-3	110–115	4.7	27.5	7000	1000	286	202	0.8
107-9-4-1	125–135	4.8	28.0	7400	1000	286	279	0.7
107-9-4-2	140–150	4.8	29.5	8200	1000	336	357	0.7
107-9-5-1	165–170	4.8	27.0	7500	1000	272	298	0.8
107-9-6-1	190–195	4.9	26.5	7200	1000	256	279	0.6
107-9-6-2	200–205	5.0	26.5	7500	1000	300	243	0.7
107-9-7-1	2105–215	4.9	23.5	6200	1000	180	215	0.7
107-9-8-1	216–280	5.6	16.0	4700	620	98	73	0.5
107-9-9-1	280–335	6.1	16.5	4900	760	86	50	0.6

[a]LBS/A = pounds per acre

glaciation were seen deeper in the section (Fig. 5. 9). Soil and sediment studies incorporated particle size, phosphorus, potassium, pH, and organic matter (OM). Textural analysis used the techniques – mechanical sieve and hydrometer – described in this chapter, and their results are shown in Tables 5.4 and 5.5.

From a lithostratigraphic perspective, Dort reduced the 13 units (9 units shown in Fig. 5.9) to a section of 4 geological units which he called: (i) Unit 1, the Bignell loess, 8000–13,000 B.P.; (ii) Unit 2, Peoria loess, 13,000–18,000 B.P.; (iii) other glacial and erosional materials; and (iv) Upper Pennsylvanian bedrock (Reagan et al. 1978). He placed the three cultural stratigraphic units into the upper two geologic units. The older of the culture stratigraphic units, "Pre-Clovis", was located on the subloess (Peoria) erosional surface implying an antiquity of at least 18,000 years. Soil chemical trends, particularly the phosphorus and potassium concentrations, as seen in Tables 5.4, 5.5, showed a correlative discontinuity at this location (65–75 cm). A similar break in the geochemical signatures was seen deeper in the section at 215 cm. Dort identified an erosional surface and probable paleosol at this location as well. No archaeological significance was attached to this lower geomorphic surface but its paleoclimatic relevance is obvious.

Earlier in this text (Chap. 4), the work of Follmer (1985) at the central Illinois archaeological Rhoads site was mentioned. Follmer, like Dort at the Shriver site in Missouri, identified a Peoria loess deposit on the till plain of the Illinoian (Pentultimate) glaciation just south of the morainic margin of the later Wisconsin glaciation. At the Rhoads site the Peoria Loess caps a sand and gravel outwash deposit termed the Henry formation (p/h), whose aggra-

Table 5.5. Mechanical analysis of selected horizons in Trench 9, Shriver Site, northwestern Missouri

Sample No	Depth (cm)	Sieve analysis			Hydrometer analysis		
		Percent passing			Percent		
		No. 10	No. 40	No. 200	<0.02	<0.002	<0.001
107-9-1-1	10–15		100	99	67.0	24.5	19.0
107-9-2-1	30–35		100	99	69.5	30.0	22.5
107-9-3-1	35–60		100	99	75.0	39.5	33.5
107-9-3-2	65–75	100	99	99	75.0	39.0	33.5
107-9-3-3	110–115		100	99	75.0	38.5	34.0
107-9-4-1	125–135		100	99	76.0	38.5	34.0
107-9-4-2	140–150		100	99	72.0	36.5	32.0
107-9-4-3			100	99	71.5	34.5	30.0
107-9-5-1	165–170		100	99	74.5	34.5	30.0
107-9-6-1	190–195		100	99	74.5	33.5	27.5
107-7-9-1	210–215	100	99	99	68.0	29.0	23.5

Analysis performed by division of materials and research, Missouri State Highway Department

dation peak was about 17,000 B.P. Follmer gives the Peoria Loess a wider time range than Dort – 12,600–23,000 years – based on radiocarbon dates for its base, near St. Louis, of 23,110 ± 280; 23,930 ± 280 B.P. (McKay 1979). On the Peoria Loess a *paleudal* or buried mollisol was identified which has the pH profile increasing with depth together with high silt (63–76%) values typical of loesses. The observed clay ratio (0.3) in the mollisol's Ap horizon increases to a value of 1.4 in the Bt2 horizon (McKay 1979). Argillic horizons in mollisols have fine-to-coarse clay ratios ranging from 1.0–1.5 (Wascher et al. 1960). Likewise, in well-drained examples, the clay mineral alteration is minimal. Follmer noted some properties, the presence of incipient E horizon, in the mollisol that indicated the possibility of polygenesis or overprinting caused by the presence of a later forest cover (Wascher et al. 1960). The C horizon is the Henry formation. Follmer does nor specifically discuss the archaeological materials found at the Rhoads site, but they are later (Archaic Period) than those excavated at the Shriver site.

6 Petrography for Archaeological Geology

6.1
Introduction

It is axiomatic that the most common artifactual material of prehistory is stone – rocks and/or minerals. Stone was the most durable of all materials available to early humans and, in most environmental settings, the most readily available. Its durability made it desirable for a multitude of tasks as well as helping ensure its survival in archaeological sites. For the archaeologist the survival of ancient human stone tools and artifacts has been both a blessing and a source of unintentional biases in terms of the reconstruction of past cultural behavior. Even with the earliest human culture surely there were other implements other than those of stone.

Another term, most familiar to earth scientists, describes the condition or process most responsible for the survival of ancient tools – stone or otherwise. That is diagenesis. This process of the chemical recrystallization or replacement in a substance, buried in the earth, given enough time, will change almost any natural material, even stone. Diagenesis might be confused with weathering which is the physical and chemical disintegration of rock and earth materials upon exposure to atmospheric agents. In the case of most rocks and minerals, the time needed for diagenetic change to occur is much longer than that which alters organic materials such as bone and botanicals. There are instances, numerous in some geological epochs, where even these latter materials become mineralized during diagenesis and are no different from stone. Happily for paleontologists, most past life forms are fossilized and preserved for study.

The archaeology of early hominids, like *Australopithecines* and the other seminal members of the genus *Homo*, can be characterized as that of a study of stones and bones, where the "bones" are mineralized fossil facsimiles of the original organic forms. In the research and interpretation of early hominid sites, the lack of anything other than stone and "bone" defeats the drawing of a more comprehensive picture of the environment and behavioral milieu of these ancient human ancestors. While an elaborate or complex material culture, expressed in a wide variety of tools, garments, and forms of shelter may not have existed, it is also as sure that the daily existence of these hominids was not depauperate as the archaeological record seems to indicate. In certain cases, such as that of the South African *Australopithecines*, the lack of primitive stone implements such as those found to the north at Olduvai Gorge led Raymond Dart to invoke the existence of an archaeologically invisible "osteodontokeratic" tool kit to provide these early hominids with "culture" (Jurmain et al. 2000). Time and careful research has helped to provide a more parsimonious picture of the South African hominids relative to that of contemporaries elsewhere in eastern Africa. In defense of the frustrated Dart, the South African *Australopithecines* and those elsewhere may have indeed used tool kits that included a large proportion of tools that were organically derived. In the millions of years since the Pliocene, only those artifacts of stone have survived in the geological, and in this case, the archaeological record.

Recognizing the presence (or absence, in the South African example) of stone artifacts has never been an archaeological failing. The *correct* geological identification has not been as common a result. Every undergraduate student in geology has been schooled in the use of a hand magnifier to help identify a particular rock or mineral. Unfortunately, for whatever reason, this has not been the case with the average archaeological student. Even for the most competent field geologist the identification of the specific form of rock or mineral is not guaranteed using such a simple inspection tool, but it is vastly superior to not venturing any informed opinion at all. In my own experience, in field and laboratory archaeology in the USA and Europe, I found that most stone finds were classified as either "chert" (in the USA) or "flint" (Europe). Those cursory and many times wrong identifications reflect both (1) a lack of recognition of the importance of a correct determination of a rock or mineral and (2) a yawning lack of cross-disciplinary training on the part of most archaeological students in America or the Old World. This chapter, in the basic level of a textbook, cannot hope to rectify a lack of methodological emphasis on earth science training on the part of students of archaeology. At best, it can suggest that both of these situations act against the best interests of archaeological inquiry and gives a brief introduction to some of the more standard ways of identifying rock and minerals.

6.2
Techniques – Optical and Otherwise

Petrology is the scientific study of rocks and minerals; petrography is the identification and description of rocks and minerals. As rocks are collections of minerals so their identification is the principal goal of petrography. The most straightforward means is as just mentioned – a 10× hand magnifier. Again, the key to one's choice of identification method is specificity. If one's goal is a simple determination of the class of rock being examined, it is reasonably sure that a distinction between sedimentary, igneous or metamorphic rock types can be accomplished with a good hand magnifier. If one wishes to ascertain a specific type of sedimentary rock – clastic, carbonate or evaporate – then this too is possible for the trained eye. Beyond this level of classification it becomes increasingly more difficult. The next level in petrographic identification requires the optical microscopes. This can be done with a low-power binocular microscope using reflected light or a more powerful, high (optical) resolution petrographic microscope used to study thin sections of rocks and minerals. Beyond the optical methods are X-ray, electron, particle-beam, and magnetic instrumental techniques of which X-ray methods are the most common (Chap. 7). In this chapter both hand specimen and thin-section examples of rocks and minerals important in archaeology are presented.

6.3
Macroscopic Analysis of Rocks and Minerals – Hand Specimen Properties

6.3.1
Color

This is one of the most obvious physical properties of lithic materials. It can be diagnostic, but as a rule it is not the major identifying characteristic as it is in the studies of soils and sediments. The reason for this is relatively simple. There are 12 soil orders while there are over 2000 minerals. Add to this fact that many archaeological rocks and minerals vary in their color (*allochromatic*), even within a specific type. Quartz has "smokey", "rose", and "amethyst" color varieties even though they are all quartz. The same thing is observed in flint or chert which has a host of varieties in different colors (cf. section later in this chapter). Some minerals, such as the metallic ones, tend to have reasonably characteristic shades of one basic color (*idiochromatic*) such as copper ores: blue (azurite), chrysocolla (blue-to-green) – and the very commonly occurring malachite is green. Chalcopyrite, a common copper ore in antiquity is yellow. Native varieties are generally more predictable – sulfur is yellow; copper is reddish-brown; slate and graphite are gray-to-black, etc. The best one can say of the property of color is that it should be used in concert with other criteria to assure confidence in identification.

6.3.2
Streak

Streak is the color of powdered minerals. This property is more reliable as an indicator of color than that of the visual color of the hand specimen. One can either rub the mineral sample on a porcelain tile or powder the sample to obtain the streak. Hematite streaks red; galena, silver-gray; hornblende, green; talc, white; azurite, blue; and magnetite, black.

6.3.3
Luster

This is the property that describes how the mineral absorbs, reflects or refracts light. Broad categories of luster are "metallic" and "nonmetallic". Pyrite and galena have a metallic luster. Quartz has a "glassy" luster as does olivine; soapstone (steatite) has a "greasy" or "resinous" luster as do many flints.

6.3.4
Cleavage and Fracture

The crystalline structure of the particular minerals is observable in their geometric shapes showing clear membership in one of the six crystal systems: cubic (pyrite), monoclinic (gypsum), triclinic (plagioclase feldspar), orthorhombic (sulfur), tetragonal (chalcopyrite) and hexagonal (apatite). A euhedral shape is one where the mineral grain has distinguishable crystal faces on all sides, such as calcite. Some mineral grains have only a few crystal faces and are subhedral while grains with no crystal faces showing in a rock are anhedral. Most minerals are subhedral as a rule. Minerals break or cleave along planes that reflect the crystal system. In general, minerals with low crystal symmetry (triclinic) exhibit distinctive cleavage, such as the feldspars. Other minerals with high symmetry are less likely to exhibit distinctive cleavage. Fracture in rocks must be distinguished from that seen in minerals. Likewise fracture is not cleavage. Cleavage planes are bonding surfaces of the crystal. Fracture can occur along cleavage planes, but it does not have to. Fracture surfaces can be nonplanar, nonparallel because a rock is a collection of minerals reflecting this heterogeneity in its overall response to breakage. A good example is the important archaeological material – flint – which does not exhibit cleavage, but rather fractures in a distinctive, conchoidal manner. Obsidian, a volcanic rock, does not have a crystalline structure and hence, no cleavage property. Obsidian does fracture conchoidally. The other common fracture forms are irregular and fibrous. An archaeological material that fractures irregularly is basalt as does most quartzite although the latter can exhibit conchoidal fracture.

6.3.5
Hardness

Hardness is a reliable property of minerals that have not been weathered or otherwise altered. Ten minerals have been arranged in a scale known as the Moh's hardness scale. The following list gives the range of hardness relative to some common minerals:

10: diamond – the hardest material; cannot be scratched
 9: corundum – can be scratched with diamond

8: topaz – can be scratched with emerald
7: quartz (also flint, olivine, garnet) – can be scratched with steel
6: feldspar and opal – can be scratched with quartz
5: apatite (also hornblende) – can be scratched with ordinary knife
4: fluorite (also dolomite, many copper ores)– can be scratched with glass
3: calcite (also mica, halite) – can be scratched with common wire nail or a penny
2: gypsum – can be scratched with the fingernail
1: talc – can be crushed between the fingers

6.3.6
Twinning

This property is observed in a host of minerals in several colloquially named forms such as the "fairy cross", seen in staurolite and the "fish-tail" observed in some gypsum and the "dove-tail" in calcite. Other commonly used names for twinned minerals are "albite twins"; "penetration twins"; "iron cross"; "spinel twins" and "star twins". This short list does not show the existence of the varieties of twins that are seen across a large number of common minerals such as calcite, feldspar, galena, gypsum, pyrite, quartz, rutile, spinel, sphene/titanite, stibnite to wurtzite. Of the silicate minerals, all of the feldspar group – albite to sanidine – form twins. Calcite is distinguished by twinning from its close mineralogical cousin, dolomite.

Twinning is observed at the hand specimen level and seen under the petrographic microscope where it is used, in the latter case, to identify minerals in thin sections. Twins result from errors in crystal growth. These errors result in characteristic twin forms and follow what are called "twin laws" – the Spinel law; the Albite law; the Carlsbad law, etc. Orthoclase twins observed according to the Carlsbad law are shown in Fig. 6.1. Microcline – a potassium feldspar with some sodium – is recognizable by its characteristic "cross-hatch" or plaid-like twinning which follows the Albite law.

6.4
Optical Properties of Minerals

While a detailed treatment is outside the scope of this introductory level, a discussion of rocks and minerals, optical properties and indices the experienced petrographer uses in identifying minerals in thin sections are included in our review of archaeologically interesting rocks and minerals. Petrographic thin sections are just that – paper-thin 30-μm slices of a rock sample that are mounted and polished on glass slides. These slides are then examined using the petrographic microscope invented in 1828 by William Nichol (Ford 1918). The use of thin sections was pioneered by Henry Clifton Sorby in his examination of Jurassic limestones from Yorkshire, England in 1851 (Adams et al. 1997).

Plagioclase

Fig. 6.1. Plagioclase feldspar in thin section. Note pronounced twinning in XPL. (Used with the permission of the University of North Carolina Department of Geoscience)

0.25 mm

The petrographic microscope is distinguished from other types of optical microscopes by its reliance on polarized light. Polarized light is light that vibrates in one specific plane. Historically, petrographic microscopes use a Nicol prism (so called for William Nicol) made of calcite to produce light that vibrates in only one plane (plane polarized light or PPL). Modern petrographic microscopes use an organic film as a filter to produce PPL.

PPL is produced by one polarizing filter located below the sample stage. A second moveable polarizing filter, with vibration direction set perpendicular to the lower polarizing filter is located above the sample stage. The light ray emerging from the lower filter vibrates perpendicular to the upper filter's vibration direction. With both filters in the light path we observe the property of extinction. This arrangement is termed cross-polars or XPL where the lower filter is the *polarizer* and the upper filter is the *analyzer*. When a thin section of a mineral or a rock with a variety of minerals is placed between the two crossed polars, one can observe various intensities and colors because the light is passing through an anisotropic material, i.e., a carbonate that bends the PPL such that not all the light is extinguished by the upper polar. The resultant light is called *birefringence*.

Birefringence is a diagnostic property of anisotropic minerals – it is not seen in isotropic fluorite, garnet, spinal or amorphous (glass – man made or volcanic) minerals – those materials in which the light ray travels unaffected through the material.

The index of refraction and birefringence are consistent optical properties of minerals used by petrographers. The law of refraction, called Snell's Law (see Chap. 7, XRD) after its discoverer, in 1621, Willebrod Snellius, who proposed its familiar form:

$$\sin i / \sin r = n \qquad \text{SNELL'S LAW of REFRACTION}$$

where i is the angle of incidence for the light beam and r the angle of refraction as the light beam exits the material and n the index of refraction or geometric ratio of i and r. Values of n fall between 1.4 and 2.0 for most minerals. Typical values are: apatite, 1.624–1.667; calcite, 1.486–1.740; fluorite, 1.434; gypsum, 1.519–1.531; olivine, 1.63–1.88; quartz, 1.544–1.553 and zircon (1.923–2.015). A few transparent minerals like diamond (2.418) and rutile (2.605–2.901) exceed the typical range for the index of refraction.

The difference in the highest and lowest indices of refraction in a mineral results in its birefringence. In petrography it is termed weak, moderate, strong, very strong and extreme.

Using PPL alone, it is generally possible for an experienced observer to estimate the refractive index and make distinctions between different birefringent crystals.

Another important aspect of a mineral's color is pleochroism. This optical property is observed in a few common crystalline minerals, biotite and hornblende being examples. When the petrographic microscope's stage is rotated in PPL, anisotropic minerals will demonstrate this property by changing colors. Biotite mica is a common pleochroic mineral and is easily identified by this property coupled with its high birefringence. Pleochroism is easily observed at the hand specimen level in transparent varieties of cordierite and spodumene.

A final optical property on grain mounts sometimes mentioned in the identification of minerals is only observed in thin sections is relief. Relief, like the properties of birefringence and dispersion, is directly related to the refractive index. Here the difference is not between the path difference or wavelength of the light in the mineral, but the difference between the mineral on the thin section and its mounting medium – epoxy, Canadian balsam, etc. Most mounting media have a well-known refractive index of 1.537 (1.54) so that minerals with a higher refractive index than 1.54 will demonstrate what is termed 'positive' or moderate to-high relief where the mineral appears bolder than media in PPL. Low or negative relief would be where the mineral appears less distinct than the mounting media. Figure 6.2 illustrates positive relief in quartz. The refractive index is very useful in determining the identity of a mineral, particularly those isotropic varieties like fluorite. A general rule among petrographers is that felsic minerals – quartz and feldspar – typically show low relief and are almost colorless. By comparison, ferromagnesium (mafic) minerals exhibit high relief and color, e.g., olivine.

In the following sections on rocks and minerals, where pertinent, their general – hand specimen or grain mount – and thin section properties will be briefly noted. The foregoing discussion of optical properties like that of the macroscopic indices is meant to point out those that are most distinctive and descriptive when used for identification. As a final note relative to identification of rocks or minerals found in archaeological contexts, a petrographer friend of mine emphasized the importance of the hand specimen in field studies. The trained eye and a good hand lens can go far in determining to a high degree of confidence the material or materials used by the prehistoric culture under study.

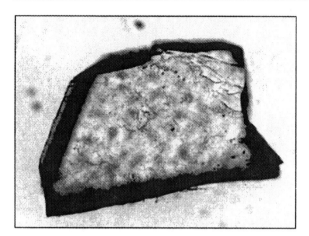

Fig. 6.2. Relief in a garnet crystal. (Modified from D. Perkins and K.R. Henke 2000, *Minerals in Thin Section*, Prentice Hall)

6.4.1   Reflected Light Microscopy

Many minerals particularly the metallic ones do not transmit light no matter how thinly they are ground. To examine such opaque minerals in polished section, plane polarized light is reflected off the upper, polished surface. Many research grade petrographic microscopes are equipped with both transmitted and reflected optics. One can differentiate between isotropic and anisotropic minerals under reflected light. Most isotropic minerals will remain an unchanging dark color as the microscope's stage is rotated. Anisotropic minerals, while never going completely to extinction, will lighten and darken four times as the stage is rotated through 360°.

6.5
Major Rock Types and Archaeology

The three rock types, sedimentary, igneous and metamorphic, appear throughout prehistory as the materials for simple domestic tools to major masonry construction. In this section most important archaeological rocks and minerals will be discussed with some illustrative examples from archaeology as well as a listing of their key petrographic properties that can be used in their identification using optical methods. We begin with sedimentary rocks (Garrells and Mackensie 1971; Pettijohn et al. 1975; Blatt 1982; Flügel 1982).

6.6
Sedimentary Rocks

Sedimentary rocks are formed by the consolidation of sediments. In the rubric of geology they are termed the "soft rocks", although the hardness of some bioclastic limestones, as we shall see later in this chapter, is formidable enough to be used as a grinding stone. Of the many sedimentary classification systems or schemes most can be reduced to three groups of rocks that are principally clastic, biogenic and chemical (Andrews et al. 1997a), which in turn can be termed clastic, carbonate or evaporites. Another way of viewing these rocks is simply *textural*. Clastic texture refers to the mineral fragments and rock debris which comprise the material. Nonclastic texture is that produced by the chemical or biogenic processes which form the rock. Within the nonclastic forms are the textures – crystalline, skeletal and oolitic. This discussion will follow the former classification terminology of clastic, chemical and biogenic (Table 6.1).

Table 6.1. Classification of sedimentary rocks

Texture	Composition	Rock name
A. Clastic rocks		
Coarse-grained (over 2 mm)	Rounded fragments of any rock type – quartz, quartzite chert dominant	Conglomerate
	Angular fragments of any rock type – quartz, quartzite, chert dominant	Breccia
Medium-grained (1/16 mm–2 mm)	Quartz with minor accessory minerals	Quartz sandstone
	Quartz with at least 25% feldspar	Arkose
	Quartz, rock fragments, and considerable clay	Graywacke
Fine-grained (1/256–1/16 mm)	Quartz and clay minerals	Siltstone
Very fine-grained (<1/256 mm)	Quartz and clay minerals	Shale
B. Chemical and biogenic (some of them are not chemical but biological)		
Medium- to coarse-grained	Calcite ($CaCO_3$)	Crystalline limestone
Microcrystalline, conchoidal fracture		Micrite
Aggregates of oolites		Oolitic limestone
Fossils and fossil fragments loosely cemented		Coquina
Abundant fossils in calcareous matrix		Fossiliferous limestones
Shells of microscopic organisms, clay-soft		Chalk
Banded calcite		Travertine
Textures are similar to those in limestone	Dolomite $CaMg(CO_3)_2$	Dolomite, dolostone
Cryptocrystalline, dense	Chalcedony (SiO_2)	Chert, etc.
Fine to coarse crystalline	Gypsum, ($CaSO_4 \cdot 2H_2O$)	Gypsum
Fine to coarse crystalline	Halite (NaCl)	Rock salt

Fig. 6.3. A banded sandstone (scale is 5 cm)

6.6.1
Clastic Rocks

Fragments of rocks are called *clasts*. Clasts are the product of the weathering, physical or chemical, of rocks. Clastic rocks are classified on the basis of grain size ranging from gravel-sized fractions in conglomerates and breccias through sandstones, siltstones to the fine-grained shales. The grain-size scale for clastic rocks is exactly that for sediments.

6.6.2
Sandstones

Of the clastic rocks the sandstones are familiar archaeologically as building stone. They are the easiest to recognize as they are cemented sand grains as their name so aptly implies (Fig. 6.3). One scarcely needs the microscope to determine this class of rocks. Within the class of sandstones are (1) the quartz sandstones, composed mostly of quartz; (2) arkose, made up of feldspars rather than quartz; (3) graywackes which contain 15% finer-grained materials – clays and silts.

6.7
Arenite or Arenitic Sandstone

General properties: While the name "arenite" is a bit dated, simply implying a quartz-rich sandstone, it still commonly appears in texts. Similar terms, not used in this chapter, are "rudite" or a conglomerate or gravel-rich stone and "lutite" – a clay-rich stone (claystone). Of the general class of "sandstones", quartz arenites are greater than 95% quartz. The grain matrix is well-cemented. Orthoquartzite is an arenite cemented with quartz. Other arenites may be cemented with calcitic, dolomitic, clayey or ferruginous material. As a rule these hard, tough rocks are well-suited to archaeological tool use.

Thin-Section properties: The grains are closely bound and relief is contingent on orientation. In PPL, single crystals are rarely euhedral. In XPL, quartz grains exhibit classic undulatory extinction. Interference colors are low order, but bright.

6.8
Arkose or Arkosic Sandstone

6.8.1
General Properties

Arkoses are defined by the abundant presence of feldspar (>25%). The feldspar may be either plagioclase or orthoclase. Arkoses are considered to be more reflective of source environments than depositional processes (Williams et al. 1955).

6.8.2
Thin Section Properties

Feldspars appear brown in PPL, colorless otherwise (e.g., not pleochroic). In XPL the twinning of the feldspars can be observed (Fig. 6.2). The "Carlsbad twin" is common in alkali feldspars, but is seen in plagioclase as well. Fine parallel twinning ("albite") is seen in the anorthoclase and microcline varieties. Relief is low or negative in both alkali and sodium/calcium feldspars (oligoclase, anorthite).

6.9
Lithic Arenite Sandstone

6.9.1
General Properties

These sandstones are quartzitic and contain more rock fragments than feldspar grains. The lithic material may be from any of the three rock types. If the rock fragments were larger than sand grains these would be called microbreccias. Depending on the rock types, these arenites can vary from immature (high percentage of feldspar) to comparatively mature (higher percentage of quartz and secondary minerals – hornblende, tourmaline, etc.).

6.9.2
Thin Section Properties

The quartz and feldspars will behave as in the arkoses and arenites while the lithic materials will display optical characteristics of these rocks and their constituent minerals.

6.10
Graywackes or Clay-Rich Sandstone ("Lutitic")

6.10.1
General Properties

These are sandstones with greater than 15% silts and clays as a matrix. Like the arenites, the graywackes vary in name according to the dominant grain types – quartzitic, feldspathic and lithic. These sandstones are generally darker in color and poorly sorted. The argillitic grains in the matrix are not visible to either the eye or under the optical microscope due to their small size (a few microns).

6.10.2
Thin Section Properties

A typical graywacke under PPL shows an opaque fine-grained matrix in which the larger quartz, feldspar, accessory mineral grains and rock fragments are seen. As with other sandstones, the optical properties of quartz and feldspars in PPL show coloration and shape according to their specific mineralogy. The XPL views show that extinction and twinning are apparent in quartz and feldspars. Rock fragments and clays will show coloration in PPL that are indicative, such as brown-colored iron oxides in the cement; pale green chlorite (black in XPL); glauconite, a brownish-green. In XPL, amphibole will show an orange interference color. Lath-like mica and particularly muscovite will show birefringence colors (Fig. 6.4) – blue-red, orange, much like the clays.

6.11
Siltstone, Mudstone, Claystone and Shale

6.11.1
General Properties

These clastic sedimentary rocks are composed of silicate minerals, mainly quartz and clays less than 5 μm in diameter. Interestingly, they are the most abundant of all sedimentary rocks, but compared to the harder sandstones

Fig. 6.4. a Muscovite mica (scale is 5 cm). **b** Extinction (*upper image*) and birefringence (*lower image*) in muscovite. (Images in **b** used with the permission of the University of North Carolina Department of Geoscience)

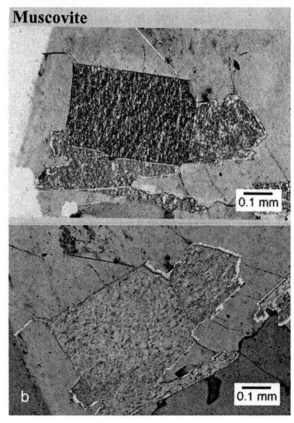

and limestones have less archaeological importance than that of a geological nature with the exception of shale (Fig. 6.5). In grainstones and mudstones the former variety is composed of calcite grains cemented by spar. Mudstone is pretty much as the name implies – consisting mostly of small silt-sized grains. The sediment is not overly compacted and the grain structure or shapes in grainstones are oolithic to peloidal. Bioclasts can occur as well.

Fig. 6.5. Shale (scale is 5 cm)

Mudstone, also known as claystone, is most commonly named shale. Because of their habit of fracture along bedding planes, shales have been used from tabular artifacts to roofing material in the past. For structural purposes, even shale is too fragile for building proper.

6.11.2
Thin Section Properties

Because of the small grain sizes in these rocks, petrographic microscope techniques can do little more than observe small-scale sedimentary structure and describe larger-grained siliclastic inclusions. PPL and XPL will show the presence of clay minerals – birefringence, etc. – and micas, such as chlorite, and muscovite. In XPL, rotation of the thin section will often demonstrate extinction. Also in PPL and XPL, the differing optical properties of adjacent laminae will appear as lighter and darker bands. The authigenic crystals of calcite, dolomite, siderite, etc., will appear highly birefringent as a rule.

6.12
Breccias and Conglomerates

6.12.1
General Properties

If one considers grain size as a continuum, clasts offer clues to the source of the sedimentary rocks – shales form from micron-sized minerals; sandstones from millimeter-sized minerals. Breccias and conglomerates are composed of the largest clasts. Morphologically, the breccias and conglomerates are differentiated on the basis of surface roundness; the latter the more rounded than the more angular breccias (Fig. 6.6). The lack of angularity in conglomerates

Fig. 6.6. Conglomerate:
a hand specimen (*scale divisions* = 1 cm), **b** Naqada stone bowl, pre-dynastic period, Egypt, made from "breccia" (conglomerate) (Plate XXXVI, fig. 60, *Prehistoric Egypt*, W.M. Flinders Petrie, British School of Archaeology in Egypt, London, 1920)

a

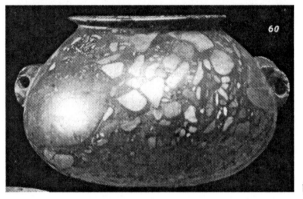

b

is considered to be the result of fluvial action. Both terms for these rock types denote their structure and not their composition. As one would expect, their mineralogical make-up can be and is quite varied, ranging from carbonate to volcanic in composition.

The identification of the breccias and conglomerates rarely requires more than a hand lens, due to the size of the fragments or pebbles. The identification of the binding matrix of the specific rock type does require a hand lens and, in many cases, recourse to microscopes. This is due to the different depositional environments in which the rocks originate. If the rock was deposited in a marine or lacustrine setting, then the matrix can be calcareous and clay-rich together with significant amounts of sand. If the rock is more the result of fluvial action, the matrix will reflect this with interstitial silts and clays bound by carbonate cements. Their matrices can vary from limonitic, calcareous, siliceous to lithified clay-rich cements. Breccias and conglomerates are important rocks from an archaeological standpoint, with their use in both artifacts and buildings (Lucas and Harris 1962; Mannoni and Mannoni 1984:159; Winkler 1994; Mazeran 1995).

6.12.2
Thin Section Properties

An example of this type of study is one by Mazeran (1995) wherein four varieties of breccias and conglomerates were identified and characterized relative to use in Gallo-Roman architecture in southeastern France. Within this suite of materials was a range typical of the variety rightly imputed to these rocks. The "breccia" of Vimines used in Roman decorative plaques and columns has a carbonate cement of a micritic/microsparite nature. The "Bordeaux Conglomerate", also classified as a "pudding stone", is held together by a molassic-glauconitic-cement of strictly marine origin. It is termed "breccia frutticulosa" by Italian archaeologists. In PPL the grains of the cement showed a characteristic green of glauconite and in XPL, weak birefringence. The cement was rich in calcite that is given to bright, high order colors. Oxidation of ovoid glauconite grains will produce brown margins because of their ferrous iron content.

Tectonic breccia, such as that of Trets-Pourcieux (France), is a metamorphosed rock. The metamorphism of this calcarenite shows features such as stylolitic ("tooth-edged") hematite crystals and mylonization (Mazeran 1995). The breccias of the French Maritime Alps are not metamorphic and are calcareous and dolomitic avalanche breccias. The matrix is a calcaro-argilliceous cement rich in hematite. The metamorphosed Italian breccias ("Dorata", "Giallo Broccatello", "Rossa") have been studied petrographically with their provenance near Sienna (Bruno and Lazzarini 1995). The Breccia Rossa Appennica in thin sections shows its marine origin with abundant pelagic bivalves, radiolarites and brachiopods embedded in a dark-brown micritic cement, hence its name. Breccia Dorata, a favorite for Roman monuments, has clasts of microcrystalline calcite with quartz cemented with calcite. The cement varies from golden-yellow to pink.

6.13
Carbonates: Limestones and Dolomites

Of the carbonate or calcareous sedimentary rock, limestones are the most common. This is because their origins are most frequently in marine basins. In the ocean the shells or skeletons of small animals rain down constantly forming thick layers of carbonate ooze. These muds lithify into the skeletal limestones. As a rule the limestones, skeletal and otherwise, are authigenic in nature, i.e., they have formed *in place* at that place of deposition. The lime carbonates or limestones are chiefly made up of calcite ($CaCO_3$) and dolomite ($Ca(Mg)CO_3$) and ankerite $Ca(Mg,Fe,Mn)CO_3$.

Rocks that are primarily calcite are termed limestone. Those with dolomite are dolostone or simply dolomite; and mixtures can be dolomitic-limestone. Some limestones have a large proportion of the calcium carbonate *pseudo-morph*, aragonite. In most limestones, particularly those exposed to meteoric water, the aragonite exsolves and recrystallizes to calcite. Such limestones are "old" limestones or compact limestones. Returning to the allogenic forms of limestones, these clastic rocks consist of calcite (Fig. 6.7) if either terrigenous, organic or oolithic in origin (Williams et al. 1955). Most are of marine origin. The fragments are cemented by additional carbonate that precipitates into the pore space between the calcite grains. If the number of silicate grains increase, the rock is a calcareous sandstone rather than a calcareous limestone or cal-carenite. Some limestones are made up of small (<2 mm) roundish grains that appear to be organic forms or tiny geoids. These rather unusual grains are

Fig. 6.7. Calcite. – *Above* Formation in a thermal spring, *below* hand specimen

ooids that are composed of carbonate formed around a detrital silica nucleus. If these forms are larger than 2 mm they are called pisoids or oncoids. Bioclastic limestones are made of cemented shell and skeletal fragments. These clasts can consist of molluscs, brachiopods, corals, forams to algae and worms. Peloidal limestones resemble ooids, but the grains lack nuclei and are completely micrite.

Micrite or microcrystalline calcite has a diameter of 5 μm or less (Adams et al. 1997). Sparite are calcite grains over 5 μm in size. The presence of these forms of calcite in limestone gives them their respective names – micritic or spary limestones. Micritization is the process by which clastic limestones particularly bioclastic varieties lose their original texture and form fine grained forms. Micrite is a major cement in limestones along with carbonate mud sediments. Loss of grain texture or clastic inclusions in limestones can result from processes like micritization, but also from pressure that compacts the pore space between grains.

By simply calling the most abundant carbonate rocks either limestones or dolomites belies their petrographical complexity with diverse textures from clastic to organic accretion. It is difficult to differentiate. Optically, between the two and the dye Alizarin Red S is used to do so much the same way sodium cobalt nitrate is used to distinguish the feldspars. Etching limestones with weak (10% by volume) HCl or 2 N formic acid is another technique. Calcite is more soluble than the rhombohedral dolomite which stands out in relief on either hand sections or uncovered thin sections. After extensive etching, only the noncarbonate minerals remain, such as chert, clay, feldspar and quartz.

6.14
Organic or Bioclastic Limestone

6.14.1
General Properties

Components of these rocks are diverse calcareous organic materials characterized by (1) overall particle size and shape and (2) internal wall structure of the organism. This latter quality is one aspect where the use of XPL has a significant role in thin sections – level studies of these limestones (Adams et al. 1984). The age of limestones can be relative to the percentage of aragonite if it was there at the formation of the stone. Over time, it will revert to granular calcite. In oysters and corals, both calcite and aragonite appear. The structural shape or form will remain after dissolution or recrystallization of the aragonite, unlike those forms which only secrete pure calcium carbonate whose shells or skeletons are lost.

The identification of a pseudomorph of calcite is beyond the hand lens and the microscope, unless it is an electron-based unit. Even then it is perhaps best to leave the identification to techniques such as X-ray diffraction. Fine sediments or micrite cement can fill in the voids or molds of lost organisms leaving casts.

6.14.2
Thin Section Properties

Identification of the organisms, overall porosity and cement are typically done in this section. Most examinations are done in PPL on stained specimens. As mentioned, XPL observation has advantages in examining organic structures. If the bioclasts are intermixed with sediments the crossed polars can differentiate these mineral grains.

6.15
Clastic or Calcarenite Limestones

6.15.1
General Properties

Made up of calcite fragments, these rocks are cemented by micrite or sparite. Sorting of the composite grains is just as with sandstones. The particles of calcarenites are either organic fragments, ooliths or carbonate rock fragments. Here, the grains are predominantly carbonate rather than silica. Nearshore beach deposits and dune materials form calcarenites. These deposits are rapidly cemented into rock by additional carbonate chemically precipitated in the pore spaces. Cementation can form rapidly what is termed "beach rock." In areas of multiple transgressive-regressive cycles, micritization and dolomitization can both cement these carbonate rocks under submarine and subaerial conditions. Hand lens examination of calcarenite samples can lead to misidentification as bioclastic sandstones. In some instances, the difference is truly one of degree so the difficulty is real. Figure 6.8 is a view of a bioclastic limestone quarry used for the manufacture of millstones in Roman Switzerland (Anderson et al. 1999).

6.15.2
Thin Section Properties

Grain size and texture are the first parameters easily distinguished in PPL. Porosity varies widely from tightly welded to poorly sorted varieties. In XPL, where the calcarenite grains are tightly packed and cement overgrowth has occurred, both the overgrowth and clasts will appear dark due to a uniform extinction color for grains and cement. Micritic cement can appear greenish-brown in PPL, while sparite is colorless. If dolomitization has affected the rock, then rhombs of this mineral can be seen in PPL or XPL. Recall our previous discussion on the distinctive optical properties of the carbonates – high birefringence, sharp relief, birefringence – can be used to characterize the limestones.

Fig. 6.8. View of millstone quarry (bioclastic limestone), Chables, Canton of Fribourg, Switzerland. Distance between fiduciary crosses is 2 m. (Used with the kind permission of the Direction de l'instruction publique et des affaires culturelles, Service Archéologique Cantonal, Canton de Fribourg)

6.16
Aphanitic Limestones

6.16.1
General Properties

Fine-grained by definition, the grains are difficult, if not impossible to see with the unaided eye or even a hand lens. Color is generally white-gray although the presence of iron or organic carbon can create pink, red and gray-black varieties. Fracture is chert-like, i.e., almost conchoidal due to the very fine grain of the stone. Fossils are not uncommon in these rocks. Karstic terranes are often aphanitic limestones. Most limestones do not make good stone tools due to their inherent low hardness. Exceptions, such as the Roman mill stones, do occur in the archaeological records.

6.16.2
Thin Section Properties

Fine-grained thin sections are dark in PPL and increased argilliceous content will enhance the XPL image – birefringence, etc. Where clays are concentrated in laminae the section will reflect this texture as well. Intercalated calcite veinlets will give the section a web-like appearance.

6.17
Dolomite and Dolomitic Limestones

$Ca(Mg)CO_3$ is a major component of limestone, although of a secondary origin, replacing preexisting carbonate minerals. Unless the euhedral rhomboidal dolomite can be seen microscopically (Fig. 6.9), it can only be

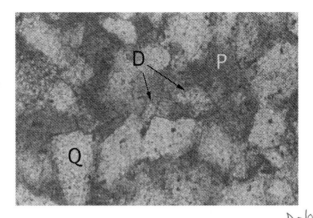

Fig. 6.9. Dolomite rhombs (*D*) in a Jurassic aged North Sea sandstone thin section. Pore spaces have been impregnated with epoxy (*dark*). Immature quartz (*Q*) grains and secondary porosity (*P*) developed from the dissolution of feldspar. Field of view is 0.23 mm. (Courtesy of Professor Paul Schroeder)

distinguished by staining after etching of the thin sections with Alizarin Red. The presence of dolomite in a limestone does not make it a dolomite. If the percentage is less than 10% it is limestone; less than 50%, the rock is dolomitic limestone, and only when 50% or more dolomite is present is the rock strictly termed a dolomite. Dolomitization is the replacement process that limestones undergo. As dolomitization is not observed in today's oceans, by the logical extension of the uniformitarianism principle, it is hard to visualize it as a primary process in earlier times. Whether a direct chemical precipitate or diagenetic replacement mineral, dolomites and dolomitic limestones are more resistant and harder than calcitic carbonates. For these properties these rocks have been regularly used in building throughout antiquity.

6.17.1
General Properties

Dolomite, as a mineral, is a major component of many limestones (Adams et al. 1997). The macroscopic appearance of dolomitic limestone is generally fine-grained, although not as fine as the aphanitic forms nor as coarse as some calcarenites. In some thin sections, the dolomite preserves the original texture or structure of the calcite protolith.

6.17.2
Thin Section Properties

Two key properties distinguish dolomite from calcite – the first is the distinctive rhombohedral shape of the single dolomite crystal and the second is its nonuptake of Alizarin Red stain. The euhedral rhombs of dolomite scattered through a limestone's calcitic matrix are singularly recognizable (Fig. 6.9). Calcite, by comparison, is generally anhedral. In closely spaced dolomitic-calcite matrices, the dolomite rhombs become subhedral or even anhedral making the necessity of staining obvious. Where dolomite has replaced calcite, the grain shape is calcitic.

6.18
Cherts, Gypsum and Ironstones

These rock types, taken ensemble, make up only a small fraction of the sedimentary rocks, but from an archaeological point of view some – notably chert and gypsum – have an importance beyond that of sheer geological abundance. Others like ironstones have had economic importance in the industrial past. Because of its use for tools since great antiquity, we shall begin our review of these rocks with chert. Chert is a hard stone; very durable yet workable in the hands of artisans.

6.19
Chert/Flint

Perhaps the most pervasive of all archaeological lithic materials, these are rocks composed of silica rather than carbonate (Luedtke 1995). They form within aqueous environments and are found within limestones or interbedded with them in sections. They can occur as either primary or secondary – the first by precipitation and the second by the process of replacement of a limestone matrix (Fig. 6.10). The silica can occur in three forms – amorphous, such as opal; cryptocrystalline or microcrystalline. All these forms are found in the chert varieties. The most common of these varieties are: flint, jasper, novaculite. In cherts composed largely of fibrous chalcedony, these are called by that name. Cherts are generally some admixture of all or some of these

Fig. 6.10. Chert nodule (scale is 5 cm)

forms. "Flint", in most English-speaking contexts, is the other common name for chert. In a strict sense, it refers to gray-to-black cherts composed mostly of chalcedony and/or cryptocrystalline quartz (Williams et al. 1955). It occurs in both nodular and bedded forms. One of the most famous archaeological localities for flint is that of Grimes Graves, Norfolk, England, the site of extensive mining activities in the Neolithic Period.

6.19.1
General Properties

Much like dolomite, chert, although silica, is thought to be a metasome as well, being less a precipitate than a diagenetic production deposited in shallow seas. The nature of cherts take three forms or "habits" – chalcedony, opaline and crystalline. Most cherts are either crystalline or chalcedony as opal is less stable over time. Opaline chert is found, but it is more likely to find a chert with amorphous opal in a matrix of quartz – fibrous (chalcedony) or crystalline (chert/flint). Cherts may have abundant impurities such as clay minerals, feldspars, dolomite and in the case of the radiolarites, abundant microfossils. Conchoidal fracture is diagnostic of almost all chert (Fig. 6.11). This property plus its durability and sharp edges made this rock a favorite with almost all prehistoric cultures.

6.19.2
Thin Section Properties

As a silicate rock the crypto- to microcrystals of quartz will demonstrate the optical properties of that mineral (cf. section on sandstone). Staining with Alizarin Red identifies any calcite in the chert. Quartz replaces calcite (metasomatic) in limestones and in thin sections this will be observed as anhedral

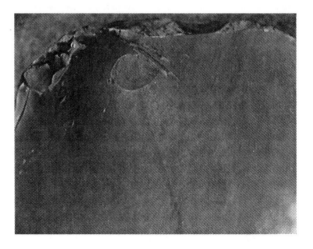

Fig. 6.11. Conchoidal fracture in chert/flint. (Photograph by author)

grains exhibiting the extinction and birefringence of quartz. Because the cherts are aggregates the extinction pattern is "patchy".

Novaculite a material used in hone or whetstones, is gray to whitish chert that is crypto- to microcrystalline, whose texture is almost granular to the touch. The true grain texture can only be observed under a petrographic microscope. Novaculite has been studied as a subject of the thermal pre-treatment of cherts to improve their working qualities for the production of flaked stone tools (Flenniken and Garrison 1975; Fig. 6.12a,b).

Jasper is a brown to red chert often difficult to differentiate from chalcedony. The reddish color originates from ferric oxides in the material. Chalcedony as seen in X-ray studies is microfibrous as a rule, but contains amorphous silica as well. This gives this chert a glossy, slick feel to the touch and in some varieties, an almost gemstone quality. Chalcedony is an opaline chert and as a result is very hard. Opal is not stable and will recrystallize over time, thus some cherts which may have been opaline have reformed into chalcedony and crystalline forms. Carnelian, a reddish-orange variety of chalcedony, was found as jewelry in fifteenth to thirteenth century Bronze Age Canaanite sites such as Hazor in modern day Israel.

Radiolarites are cherts with large proportions of the skeletal remains of radiolarians diatoms, sponge spicules in the stone matrix of quartz and opal. These are very hard rocks and were used in prehistory to make durable tools. Jurassic radiolarites are found in California's prehistoric cultural middens.

Rock Gypsum – after carbonates, the hydrous sulfate mineral, $CaSO_4 \cdot H_2O$, precipitates in the evaporation of marine basins. The production of gypsum or its chemical replacement anhydrite is temperature-dependent. Above 42 °C, gypsum will precipitate, below this temperature anhydrite-anhydrous $CaSO_4$ forms. At or near the earth's surface and on exposure to meteoritic waters, gypsum replaces anhydrite.

General properties: classified here under the sedimentary rocks, these forms are more correctly minerals. While less durable as a building stone in antiquity (ex. Knossos), alabaster has been mined and used for veneer slabs as well as for sculpture. Rock gypsum is a massive, coarse-grained, grayish-to-white rock composed almost exclusively of aggregates of gypsum. Where iron oxides are incorporated into the matrix, it is colored various shades of orange to red.

Thin section properties: gypsum, in thin sections, is readily distinguished from anhydrite. Gypsum exhibits low birefringence and negative relief compared to higher birefringence and lighter relief of anhydrite. The interference color of gypsum is green while anhydrite has bright interference colors.

Ironstone – deposited in marine contexts, this carbonate rock is more than 15% iron. The iron can be in the form of siderite ($FeCO_3$ and chamosite). The latter can oxidize to the iron oxide limonite. Limonite can appear in oolithic form. Hematite, Fe_2O_3, is more stable in terrigenous ironstones than either siderite or chamosite. Ironstone can be bioclastic and some are banded with chert layers (Adams et al. 1997).

Fig. 6.12. Novaculite.
a Two Early Archaic Period, ca. 8000–5000 B.C., spear points (replicates), **b** photomicrograph of microgranular quartz in novaculite. Field of view is roughly 1 mm

Travertine – deposited in caves, this carbonate is also known variously as "flowstone" or "dripstone". It is distinguished by its lamellar or banded nature which reflects its origin in the waters percolating into solution featured as in caves. Archaeologically it can occur as a finish stone because of its beauty, but more recently it has become the object of paleoenvironmental studies wherein the bands of the rock are equated to the "growth rings" of trees. It has been demonstrated that speleothems – stalactites and stalagmites – can be dated using techniques such as uranium-thorium series (U-Th), electron spin resonance (ESR) and radiocarbon. Coupled with the examination of stable isotopic ratios in the growth bands, it is possible to retrodict past climatic conditions (Hendy 1971; Shopov et al. 1994; Brook et al. 1996). With regard to cave sites, the presence of flowstone coating prehistoric artifacts, skeletal remains or art can reassure researchers of their veracity and relative antiquity. Examples of this are found at Petralona cave (Greece) and Cosquer and Chauvet caves in France. At Petralona cave, a skull of an archaic *Homo sapiens* was encrusted with travertine, which attested to its age geologically and geochronologically (Schwartz et al. 1980; Boris and Melentis 1991). At Cosquer and Chauvet, particularly at the former locations, the spectacular ancient cave art has calcite encrustations over some paintings and engravings (Clottes and Courtin 1996; Clottes 2001).

6.20
Igneous Rocks

Igneous rocks are formed by the solidification of molten magma or rock that has been melted in the upper mantle. This rock has been incorporated into the mantle from various sources, but in general it is the stuff of crustal rock which has been subducted at plate margins after crustal plate collisions. As such the rock to be melted can be sedimentary, metamorphic or igneous in nature. The type of rock melted determines the nature of the chemical components of the magma. Igneous rocks are mainly composed of the eight rock-forming minerals – quartz, plagioclase feldspar, K-spar or orthoclase, muscovite mica, biotite mica, amphibole, pyroxene and olivine.

The most common classification of igneous rocks is that of the Internal Union of Geological Sciences (IUGS) Subcommission of the Systematics for Igneous Rocks (Streckeisen 1979). Before the IUGS classification scheme, several others used in the study of igneous rocks produced hundreds of names. The names used in this text will be those of the IUGS system. The IUGS classification of plutonic or intrusive igneous rocks is less debated than that for volcanic or extrusive igneous rocks due to the fine grained character of these rocks. The color index – light-to-dark – is commonly used along with mineralogy and texture in describing igneous rock. In our discussion of igneous rocks we emphasize their modal mineralogical classification rather than their chemistry.

The texture of igneous rocks ranges from coarse- to fine-grained or pegmatic/phaneritic to porphyritic/aphanitic (Andrews et al. 1997b). Texture refers to the size and shape of individual mineral grains. Phaneritic texture is visible to the unaided eye, while aphanitic texture is not. The texture of an

TEXTURE
igneous rock is an indication of the cooling rate of the parent magma. Large-grained or very coarse(pegmatic) igneous textures are found in the plutonic rocks while finer-grained textures are found in volcanic rock. Obsidian is the finest "grained" of all igneous rocks being in actuality a volcanic glass. In the field an archaeological geologist can usually only evaluate color and texture if the grains are phaneritic and estimate the mineralogy of an igneous rock sample. The latter property, in its most generalized form, grades igneous rock mineralogy from felsic (feldspar-silica-rich) to mafic (ferromagnesium-rich; Table 6.2).

Igneous rocks are found in almost all areas and are either intrusive, plutonic forms or the more localized, extrusive volcanic varieties. In the former, we see more metamorphic rocks associated with the intrusive geology. Sedimentary facies can blanket more ancient igneous structures or the igneous rocks can force their way up through layers of older rock. Either process results in the presence of igneous rock that was recognized in antiquity for its durability and utility, hence its presence in archaeological deposits. In the case of obsidian, this volcanic glass became one of the earliest economic rocks.

In Mesolithic Greece, Franchthi cave on the Argolid coast has obsidian in low percentages (~5%) increasing through Neolithic times to 95% of the lithic component of the archaeological assemblage (Cherry 1988). This obsidian has a unique provenance. It was quarried on the island of Melos, 150 km distant, meaning that sea-faring watercraft in the Aegean have a great antiquity. At the Anatolian site of Catal Hüyük, which flourished between 7500 and 6500 B.C., obsidian was a prominent trade mineral being mined in the volcanic flows of the Konya region (Cunliffe 1988). The quantitative characterization relies on many of the optical techniques already described in this chapter. To the archaeological geologist examining igneous rocks the questions are mainly of archaeological provenance – economic, cultural, cosmological, etc. – rather than geological provenance, e.g., magmatic type, temperature, etc. The composition of the rock will assist the researcher in determining the cultural history of the material.

Table 6.2. Common Igneous Rocks

SiO_2, Na_2O, K_2O; felsic minerals → CaO, MgO, FeO; mafic minerals

Chemical type Texture	Acid	Intermediate Rock names		Basic	Ultrabasic
Fine-grained, extrusive volcanic lavas	Rhyolite Obsidian Pitchstone	Andesite Dacite	Basalt		
Medium- to coarse-grained	Granite		Diorite	Gabbro	Peridotite[a]
Plutonic		Granodiorite			Pyroxenite[a]

[a]Metamorphosed equivalents are soapstone and serpentine

To do this the investigator must determine the mode (mineralogical content), texture and color. The latter parameter, formalized originally by Johannsen (1938) into the color index is based on the percentages of light and dark minerals present. When and where possible the characterization should proceed to optical studies and where necessary, instrumental examination. Many of the following descriptions are based on the excellent *Atlas of Igneous Rocks and Their Textures* (MacKensie et al. 1982). Where terminology differs from that of American petrological usage, this volume will attempt to use the terms more familiar to the American student while recognizing the equal status of the more international terminology.

6.21
Granite

6.21.1
General Properties

Granites are intrusive, phaneritic rocks with a low or leucocratic color index consisting of mainly quartz, feldspars and micas (Fig. 6.13). Of the feldspars, potassium feldspar is the most prevalent form. The granites are therefore classified as *felsic.* Granite has been a principal building stone through antiquity. The Egyptians used Aswan granites in the Giza pyramids for lining material of passages, chambers and door frames (Lucas and Harris 1962). In contrast to the Egyptians, granite was little used by the classical Greeks (Higgins and Higgins 1996). Where granite cores the Greek Islands, the weathering of this rock results in fertile soils (Melas 1985).

6.21.2
Thin Section Properties

In PPL, quartz appears featureless as a rule. The feldspars will vary in the property of twinning, ranging from microcline's cross-hatched variety to simple twin patterns in potassium feldspar (Fig. 6.2). Plagioclase shows albite twinning. Accessory minerals such as the micas are recognized by their interference colors in XPL and their pleochroism in PPL.

6.22
Diorite

6.22.1
General Properties

Sometimes termed the "black granite", diorites are medium- to coarse-grained igneous rocks. Many of these intermediate color index rocks are melanocratic or dark colored, hence their common name. The mafic crystals

Fig. 6.13. Granite: **a** hand specimen, **b** detail of hand specimen showing feldspar and biotite

a

b

of pyroxene and hornblende along with biotite are mainly responsible for the coloration. Quartz and potassium feldspar, if present, occur in trace amounts. Diorite was used in sculpture by the Egyptians from the Neolithic into Pharonic times (Lucas and Harris 1962; Fig. 6.14). The cosmetic palettes, so popular with ancient Egyptians, were made from diorite as were mace or axe heads, bowls and vases (Lucas and Harris 1962). Caton-Thompson and Gardner (1934) reported diorite artifacts from the Fayum Neolithic sites as well. In Egypt, a dioritic-gneiss occurs often called diorite by early authors.

6.22.2
Thin Section Properties

In PPL, the plagioclase crystals are lath-like. Twinning is common with the single Carlsbad Twin to multiples of the albite variety. Potassium feldspar will be interstitial to the plagioclase. Interspersed through the thin section will be

Fig. 6.14. Diorite – Monumental head of Ramses II at Luxor, Egypt. (Photograph by Prof. Gilles Allard)

the brown and green crystals of the ferromagnesium minerals – hornblende (amphibole) and biotite.

6.23
Andesite

6.23.1
General Properties

Andesite is considered to be the extrusive equivalent of diorite (Andrews et al. 1997b). Most andesites are porphyritic. Colors range from gray to dark gray. The rock contains ferromagnesium minerals – pyroxene, amphibole and biotite – with feldspar phenocrysts. Finely grained, natural andesite was used for hand axes in east Africa (Charles Peters, pers. comm.). This "andesite" is only so based on its chemistry – it is not representative of most andesites formed in volcanic arc areas.

6.23.2
Thin Section Properties

The PPL and XPL views of andesite clearly demonstrate the microcrystalline ground mass of plagioclase. The ferromagnesium crystals together with larger phenocrysts "float" in the andesine (Fig. 6.15).

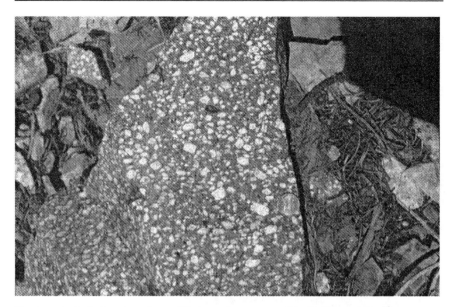

Fig. 6.15. Andesite hand specimen. (Photograph courtesy of Prof. Mike Rodin)

6.24
Basalt

6.24.1
General Properties

Basalt is an extrusive igneous rock. Basalts form at mid-ocean ridges, rifts and hotspots. In the Pacific Northwest and the Yellowstone Basin great flows of ancient basalts form impressive structures. Basalt is the most abundant fine-grained igneous rock. Basalt has a long history of human use from grinding stones, hammers, to construction stone and sculpture. In the last category, the great carved heads of the Olmec culture were made from basalt (Winkler 1994). Basalt was quarried in the Old Kingdom of Egypt in the Fayum, west of modern-day Cairo (Lucas and Harris 1962). Because this rock does not demonstrate conchoidal fracture, some archaeologists have difficulty in recognizing intentional fabrication of the stone as tools.

6.24.2
Thin Section Properties

Commonly, basalts are composed of plagioclase and ferromagnesium minerals – magnetite, olivines, pyroxenes, etc. (Fig. 6.16). However, the relative abundance of these minerals is quite variable. The glass in basalts will appear

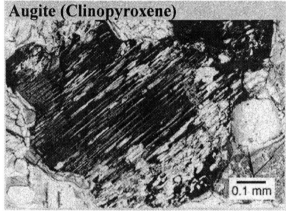

Fig. 6.16. Basalt. *Above* Hand specimen, *below* clinopyroxene crystal in thin section. (Thin section image used with the permission of the University of North Carolina Department of Geoscience)

as a dark groundmass – genuinely so in XPL – and form the back drop for the plagioclase laths and ferromagnesium crystals. The optical properties of the plagioclase have been detailed earlier in this section. By far the most colorful mineral in a basalt thin section will be the highly birefringent olivine with blues, pinks and reds prevalent.

6.25
Rhyolite and Obsidian

6.25.1
General Properties

An extrusive porphyritic volcanic rock, but unlike basalt, rhyolite is felsic in nature. Some geologists carry the definition into the category of felsic volcanic glasses making obsidian a variety of "rhyolitic" rocks. Taking this approach, the rhyolites can range from totally crystalline to totally glassy hyaline. Quartz and feldspar phenocrypts are common in fine-grained rhyolites and obsidians (Mackensie et al. 1982). In antiquity, in igneous terranes, early humans used rhyolitic rock for tools such as hand axes and crude flake implements. Obsidian, which does fracture conchoidally, was a favorite material of early lithic technologists as discussed earlier. This volcanic rock has, the sharpest edge of any cutting medium and was also recognized for its beauty (Fig. 6.17). Melrose green rhyolite, from southeast New England, was used for Middle Archaic to Late Woodland Period artifacts and has been described by Hermes et al. (2001). This pale green rock was confused by several archaeological workers before being characterized mineralogically using optical and geochemical methods (Hermes et al. 2001).

Fig. 6.17. Obsidian. *Above* Hand specimen, *below* image of Tezcatlipoca, "Smoking Mirror", whose foot was an obsidian mirror, from the Codex Fejervary-Mayer

6.25.2
Thin Section Properties

Because of their felsic nature, thin sections of these rocks can be pretty colorless. Indeed, the only birefringent crystals will be those of an accessory or accidental origin. The microcrystalline-to-glassy ground mass will be brown-to-gray in PPL and quite dark in XPL. The phenocrysts of potassium feldspar will show little twinning, while those of plagioclase will demonstrate abundant twins. The quartz phenocrysts will be of a higher relief than the groundmass, but appear clear-to-colorless in PPL and gray in XPL with some extinction.

6.26
Gabbro/Diabase

6.26.1
General Properties

An intrusive, mafic igneous rock fine-grained gabbro is found in dikes and sills is given the name "diabase" by European workers. Varieties range from olivine-rich gabbros to plagioclase-rich varieties. Gabbro contains no quartz, by definition. These rocks are coarse-grained (phaneritic) and are generally dark. The gabbros were used only rarely for building stone in antiquity because of the friable nature of the rock matrix. It was not uncommon to see this material used in smaller artifacts such as polished implements and items such as cups, bowls, plates and ceremonial items such as mace heads. Egyptian culture of prehistory used both diorite and gabbro in all these categories (Caton-Thompson and Gardner 1934).

6.26.2
Thin Section Properties

In the olivine gabbros, augite and plagioclase enclose equant and round olivines (MacKensie et al. 1982). Plagioclase generally occurs in laths, particularly in inter-granular forms, although the laths are not aligned or oriented as in trachytic gabbros where this is seen. The optical properties of the dominant plagioclase, augite and olivine are colorful under XPL. Diabases in thin sections vary from olivine-to-pyroxene-rich plates and laths of these minerals as well as ilmenite and magnetite. Calcite is often seen with its bright colors in XPL. Diabases show wide textural variation, even in thin sections, and some are notably porphyritic (Williams et al. 1955).

6.27
Tuff

6.27.1
General Properties

Tuff is an extrusive felsic-to-intermediate pyroclastic igneous rock composed of volcanic ash and a variety of mineral grains and rock fragments. Its color ranges from dull earth tones of light brown-pink to grays. Crystals and inclusions of sedimentary rock, lava fragments, and other pyroclasts are common within tuffs. Tuffs appear to have been rarely used in construction in antiquity. In ground or polished stone vessels, the use of tuff does appear in early dynastic Egypt. A very small fraction (4%) of stone vessels found in The Old Kingdom (first Dynasty) tomb of Hamaka, at Saqqara, were of "ash" or tuff (Emery 1961).

6.27.2
Thin Section Properties

Under low power magnification, the heterogeneous texture of this rock is quite apparent. Ash will appear as dark inclusions while the glass inclusion will be colorless and opaque or translucent in PPL. Welded tuffs will have a glassy ground mass, with lamination apparent (Fig. 6.18). The quartz and feldspar crystals will be elongated or flattened with little crystal shape evident except in the unwelded tuffs.

6.28
Metamorphic Rocks

Metamorphic rocks represent rock that "changed form" – recrystallized due to changes in temperature and/or pressure. These changes typically take place along plate junctions and intrusive/extrusive igneous boundaries. The most restricted form of metamorphism is that of contact metamorphism directly adjacent to intrusive igneous bodies where heat and hydrothermal fluids can cause new crystallization (neomorphism) and precipitation or replacement (metasomatism) of minerals. Most metamorphic rocks result from regional metamorphism which occurs over large areas associated with mountain building. Within mountain belts slates are formed from shales; quartzites from sandstones; gneisses formed from sediments and granites; serpentine and steatite from olivine-pyroxene-talc-rich rocks; marbles from limestones and greenstone from basalt and andesite. The time scales are long in most instances, ranging from thousands to millions of years.

Metamorphic rocks are described by texture and mineralogy. These rocks can have igneous, sedimentary or metamorphic origins and as such retain

Fig. 6.18. Welded tuff. **a** Hand specimen (*scale divisions* = 1 cm), **b** thin section showing quartz (*clear*) and biotite (*dark*) phenocrysts. (Thin section image used with the permission of the University of North Carolina Department of Geoscience)

much of the character of the original lithology while being the changed species that they are. Some of the most archaeologically important metamorphic rocks are marble, quartzite and soapstone. Upon finding such rocks away from a orogenic zone, such as on a coastal plain with nothing, but sedimentary facies for hundreds of kilometers, an archaeologist must seek cultural (trade, exchange, curation) reasons for their presence in these sites. Metamorphic rocks are classified according to their mineralogy, texture, facies or grade. In terms of their mineralogy, V.M. Goldschmidt in 1912 correctly observed that most metamorphic rocks contain only a few – four-to-five major minerals (Philpotts 1989). The mineral assemblage in a metamorphic rock helps the petrologist to interpret its thermodynamic history (Fig. 6.19). This is less important to the archaeological geologist, but the relationship between metamorphic grade – low, intermediate and high – and texture is good to know. Grade refers to temperature and pressure, with a high-grade metamorphic rock more recrystallized than a low-grade metamorphic rock.

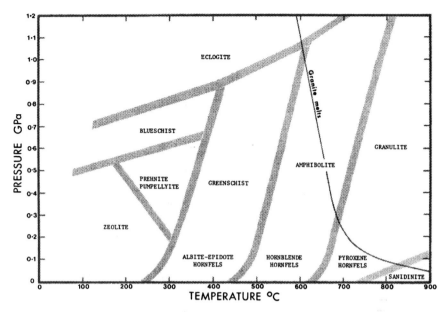

Fig. 6.19. Pressure–temperature diagram illustrating the formation conditions for various metamorphic facies. (Modified from Philpotts 1989)

Texture refers to a metamorphic rock's fabric orientation and direction or the lack of it, e.g., foliated vs. nonfoliated. The bulk of metamorphic rocks demonstrate some form of foliated texture.

6.29
Metamorphic Rocks – Foliated

Slate — a very planar, parallel foliation (Fig. 6.20).

Phyllite — foliation of fine-grained sheet-like minerals such as muscovite and biotite. Phyllites are known for a glossy luster called "phyllitic sheen" (Fig. 6.21).

Schist — fine- to coarse-grained platy minerals are aligned parallel to sub-parallel fabric which is obvious to the eye. Like phyllite, such a rock is said to have "schistosity" (Figs. 6.21, 6.23). Schists fracture along the foliations. The variety of minerals in schists is diverse, e.g., chlorite, tourmaline, muscovite, biotite, garnet, hornblende, staurolite, kyanite and sillimanite.

Gneiss — gneisses have alternating layers of different minerals such as quartz and biotite. In Europe, gneiss refers to coarse grained, mica-poor rocks irrespective of fabric (Fig. 6.22; Yardley et al. 1990).

Fig. 6.20. Slate – hand specimen (*scale divisions* = 1 cm)

Fig. 6.21. Schist – hand specimen illustrating foliation texture and phyllitic sheen. (*scale divisions* = 1 cm)

6.30
Nonfoliated Metamorphic Rocks

Hornfels   the minerals in hornfels are microcrystalline giving the rock the property of conchoidal fracture like chert and obsidian.

Granulite   this is an even-grained garnet or pyroxene-rich rock lacking the micas or amphiboles. Its essential components are orthoclase feldspar, plagioclase and quartz. This texture is observed in marble, quartzite, greenstone, serpentine/serpentinite and skarn. It is plane-foliated that can be laminated parallel to the foliation. This characteristic leads to slab-like breakage and usage in construction in antiquity.

Fig. 6.22. Gneiss. a Boulder showing characteristic banding of lighter quartz-feldspar layers and darker mica (biotite) – hornblende layers, b thin section showing quartz (*gray*), feldspar (*white*) and biotite (*black*). (Thin section courtesy of Prof. Paul Schroeder)

Amphibolite   This rock is medium- to coarse-grained. The rock is mainly hornblende and plagioclase. The parallel alignment of the hornblende crystals gives the rock its character, but seldom produces a foliation.

6.31
Metamorphic Rocks of Archaeological Interest

6.31.1
Quartzite

6.31.1.1
General Properties

This metamorphic rock generally originates as a quartz-rich sandstone facies such as an arkose or arenite. It is a product of regional metamorphism. As a rule it is a white rock weathering or staining to colors such as cream or orange in iron-rich sediments. Quartzite can appear gray when minerals such as

Fig. 6.23. Mylonite texture as seen in a typical metamorphic rock – a chlorite schist

biotite or magnetite are present with the quartz grains. Quartzite is very hard and durable, but difficult to work. As an archaeological material quartzite appears as a variety of implements, such as spear points, knives, perforators or almost any stone tool form. Because of its texture and coloration, the whiter varieties were fashioned into adornment such as beads, pendants and the like. In some of the earlier literature on the use of quartzite in antiquity such as Egypt, the rock was described as "hard sandstone" (Lucas and Harris 1962). This "quartzite", if indeed it is of metamorphic origin, was used in making sarcophagi and statuary.

6.31.1.2
Thin Section Properties

It is predominantly a fabric of interlocked quartz grains (Fig. 6.24) which show recrystallization of varying grain size. Smaller grains are seen in quartzites that have undergone more intense deformation (Philpotts 1989). The grains of quartz will show the optical properties of this mineral – low relief, little or no birefringence, colorless and some undulatory extinction in more isotropic grains. Unlike a traditional sandstone, a quartzite can show a lamellar or parallel fabric to the grains.

6.31.2
Marble

6.31.2.1
General Properties

Marble is one of, if not the most archaeologically important of the metamorphic rocks. Fine-grained varieties have been used for sculpture since antiq-

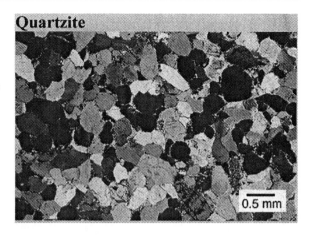

Fig. 6.24. Quartzite in thin section. Many of the grains demonstrate extinction (*black*). (Image used with the permission of the University of North Carolina Department of Geoscience)

Quartzite

0.5 mm

uity in both the Old and New Worlds. In the classical period of Greco-Roman tradition, marble became a preferred building stone and continues to enjoy wide use today as both an interior and exterior stone. In the Aegean, marble appears as a quarried stone and sculpture media in the Cycladic Islands in the third millennium B.C.E. (Cherry 1988). Cemeteries on islands such as Melos, Naxos, Paros, and Andros have produced thousands of marble figurines and vessels. In the case of Melos, the marble must come from adjacent islands as it has none of its own. As we have seen, Melos was an early (Neolithic) source for obsidian.

In Egypt, marble was used in the Neolithic for bowls and vases (Mannoni and Mannoni 1984). By the time of the Old Kingdom, at least at Saqqara site of the first pyramid – the so-called Step Pyramid – a royal tomb was constructed of calcareous marble in the twenty-eighth century B.C. In Archaic Period Greece (ca. seventh century B.C.), the use of marble for statuary increases with its use in life-size, and greater than-life size, representations of young men (*Kurai*) and young women (*Kurii*; Fig. 6.25). Into the Classical and through the Hellenistic Periods of Greece the demand for marble for both building and sculpture opens quarries across the Mediterranean adding to those of Egypt and Greece, with sites in Anatolia, Sicily, and Italy, most notable Carrara.

Marble is a granular rock, composed principally of calcite and is derived from limestones. Of course, this oversimplifies the diversity encountered in marble texture and coloration which, in the latter, ranges from the white of Greece's Pentelic facies to the gray marbles of Vermont (USA) and pink-to-red marbles of Georgia and Alabama, USA, and Carrara, Italy. Richly colored banded or "streaked" varieties are found in Italy and elsewhere and doubly metamorphosed brecciated varieties in Carrara. The color and textual variety

Marble

0.25 mm

Fig. 6.25. Marble. **a** Almond-eyed *Kore* in the Parthenon Museum, Athens, ca. 500 B.C., inventory no. 674, **b** thin section view of the interlocked calcite grains. (Photograph by the author; image in **b** used with the permission of the University of North Carolina Department of Geoscience)

of those marbles is the result of their precursor minerals and their metamorphism. The marbles of peninsular Italy are generally formed at depth by regional metamorphic processes while those of the Aegean islands can be the result of contact metamorphism. The white marbles, i.e., Pentelic, are almost 100% calcite while impurities such as quartz, carbon, clay and iron give color to others – quartz, clay, iron produce yellow or pink coloring; carbon and other organics produce gray to black marbles; greenish clay-rich marls, micaceous minerals and serpentines can produce gray-to-green colors. Marble colored by its component minerals is termed idiochromatic and those by inclusion pigments in cements between grains or in veins and interstices are called allochromatic.

6.31.3
Serpentinite and Steatite

6.31.3.1
General Properties

Serpentinite is a low-grade metamorphic rock originating from recrystallization of mafic minerals such as olivine and pyroxene (Williams et al. 1955). Olivine in hydrothermal fluids reacts to form serpentine at low temperatures in the near-surface (Yardley et al. 1990). Some European texts (Monttana et al. 1978) as well as most early archaeological workers have had difficulty differentiating serpentine, the ultramafic mineral from the metamorphic rock, serpentinite. Indeed, the terms serpentine and serpentinite appear interchangeably for the description of the metamorphic rock. Brecciated rock such as the ophicarbonates of northern Greece (Thessaly) produced from peridotites, mistaken or missnamed as "marbles", give interesting textural and color patterns (Mannoni and Mannoni 1984). These rocks are serpentines with veins of carbonates.

In the USA and Canada, steatite is a talc or "olivine-talc-carbonate rock" low in serpentinite and the accessory talc which grades into high talc (~10%) ultramafic metamorphic rock. Steatite is also known as "soapstone" in the New World. It is also called "potstone" in the Old World. The latter term reflects its use in Stone Age cultures as a vessel material. In prehistoric North America soapstone/steatite vessels appear in middle to late Archaic Period cultures (5000–1000 B.C.). In the Appalachians and in the prehistoric sites of the Chumash and Gabrialeno cultures of California, soapstone was used in vessel manufacture. Steatite vessels are a diagnostic artifact of the site of Tepe Yalya found intrasite and as trade goods across Mesopotamia (Lamberg-Karlovsky 1974). The famous pipestone steatite quarries of eastern South Dakota were used by American Indians to fashion elegant "pipestone" pipes for tobacco. In many steatites, chlorite appears as an important accessory mineral. The color of serpentinite and steatite generally runs from dark green to lighter gray-green varieties due, in the latter case, to talc and chlorite.

6.31.3.2
Thin Section Properties

The serpentine crystals of serpentinite and steatite are generally the replacement of pre-existing olivines and pyroxenes, hence the fabric may reflect this in PPL. The texture is lamellar in most cases with crystals of pyrite, magnetite and re-grown olivine sometimes present. Some serpentinites are completely recrystallized with no trace of relic grains. In XPL, these are quickly differentiated from the dark serpentine due to their relief, birefringence colors and other optical properties. The talc in steatite will be especially birefringent showing bright yellow and other colors reminiscent of muscovite, as mentioned earlier.

6.31.4
"Greenstone"

6.31.4.1
General Properties

Defined as any altered basic igneous rock which is green in color due to the presence of chlorite, actinolite or epidote (Parker 1994), this rock is metamorphic and as a category is used by archaeologists to describe stone artifacts such as axes. The coloration and texture give the material its name rather than a strict reading of its mineralogical makeup. Actinolite schists have both the desired color – green to almost black – plus the hardness that a tool or implement must have. The rock polishes into what is termed "groundstone". It is its hardness and toughness which make this material so attractive to human groups and dominate the way it is worked. While it takes a moderately good edge, it withstands shock well and so can be used for artifacts which might shatter during use. In the late Neolithic of Europe, in Switzerland and neighboring lands – Austria, France, etc., – both solid and shaft-hole axes are quite common. These axes are made from a variety of "greenstone" or "roche verte" (Ramseyer 1992). The shaft hole axes are also known as "battle axes" or "hammer axes" (Briad 1979; Champion et al. 1984; Fig. 6.26). Also termed "jadeite" which produces some confusion as jadeite per se is mineralogically a pyroxene. What we can say is the axes are uniformly made of a green "rock" – light to dark – that takes a good polish. Swiss geologists identify polished stone materials ranging from serpentine, quartzite to diorite (Winiger 1981). Seen in western Switzerland and elsewhere in the Alps as a suite of metamorphosed oceanic/pelagic sediments within which the serpentinized peridotites occur (Hsü 1995). During subsequent glaciations of the Pleistocene, the ophiolitic rocks were deposited along the morainic margins of the Rhône glacier and readily available for the Neolithic artisans.

Elsewhere, ophiolites occur along other subduction zones such as that of the Aegean basin. An example is the use of metamorphic rock, in particular jade, in which, as we have seen, either jadeite or nephrite is the dominant

Fig. 6.26. "Battle axe" with handle recovered in 1999 by the Archaeological Service of the Canton of Zug, Switzerland. Axe is made of serpentinite. The handle is 120 cm in length

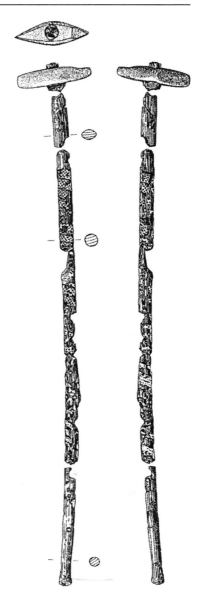

mineral (Higgins and Higgins 1996). On the Cycladic island of Syros, serpentinite and jadeite rocks occur in quantity and were used for axe manufacture. "Greenstone" (nephrite and bowenite) was a major resource for the human occupants of New Zealand prior to contact with Europeans. The sources for this material are found in a number of locations within the South Island.

Given the disparity in the identification of archaeological "greenstone", it is likely the choice of specific material was simply made to fit local lithic resources and we, as archaeological geologists, apply our specificity to the individual case. As such, the general and optical properties of the rock will follow that of the specific mineralogy involved. For instance, the "roche verte" of western Switzerland's perforated "battle axes" is clearly metamorphic and in most cases the material is either a schist – specifically a schist rich in actinolite, chlorite and/or epidote – or a serpentinite. The texture of the former is rough or raw rock is schistose where, in the finished artifact, the polished stone is dominated by the color of the actinolite and limonite grains.

6.31.4.2
Thin Section Properties

In thin sections the amphibole actinolite appears green while the quartz and feldspar is granoblastic with attendant optical properties of these minerals. Actinolite grains will show a pronounced pleochroism. In XPL, the epidote will show bright birefringence colors. As we have previously said, serpentinite – a greenschist facies rock as well – is fine grained with low birefringence. In PPL jadeite crystals demonstrate a high relief and they typically appear as platy and interlocked. Under XPL, the jadeite grains show a low birefringence.

6.31.5
Slate

6.31.5.1
General Properties

A regional metamorphic rock of pelitic or pelitic-arenaceous origin, slate's most recognized precursor is shale. It is fine-grained and demonstrates a planar fabric that gives rise to the definition of slatey foliation. The slates are uniformly gray-to-black and the principal components of the fine-grained ground mass are chlorite, muscovite, quartz and hematite (Yardley et al. 1990). Other minerals are andalusite and cordierite which can form elongated crystals (porphyroblasts) as do tourmaline, rutile and pyrites (Williams et al. 1955). Graphite represents the lowest grade of metamorphism. This rock is found throughout antiquity with its usage as palettes in Egypt, for flooring and roofing, and for its most ancient archaeological appearance as a slate "tablet" for Paleolithic etchings of mammoths and figures at Gönnersdorf (Bosinski 1992).

6.31.5.2
Thin Section Properties

In thin sections, the fine-grained layering of the parent facies is often preserved in the groundmass. XPL illuminates the micaceous minerals present

10.10

with high birefringent colors. Porphyroblasts will demonstrate the appropriate optical properties for the specific mineral; in reflected light pyrites appear yellowish-white in PPL; rutile with red-brown pleochroic colors in XPL; and chlorite with high relief, more brightly colored in PPL, but less so in XPL.

6.31.6
Gneisses and Schists

6.31.6.1
General Properties

The varieties of these rocks are usually extensive related to mineralogy. Grain size is medium to coarse in both types with augen gneisses having a porphyroblastic texture. Most gneisses appear lineated or banded with alternating quartz/feldspar and amphibole/biotite. Schists are foliated.

Figure 6.27 shows a petroglyphic gneiss boulder found in Forsyth County, Georgia, roughly 3×1 m in size. On its upper portions, it is covered with concentric circle motif petroglyphs of unknown age. It was first described by C.C. Jones in 1873 in his *Antiquities of the Southern Indians*. It has been suggested that the stone originally stood upright in the manner of a *menhir* with the "upper" end (left in the picture) being an effigy.

Fig. 6.27. Gneiss – a petroglyphic boulder from Forsyth County, Georgia. It was first described in 1873. The *lower portion* to the right of the boulder is covered with concentric circles while the *upper portion* is thought to be an effigy of an owl. Probably from the Woodland Period, ca. 1000 B.C.–1000 A.D. in the American southeast. Length 2.7 m

Fig. 6.28. Anthropomorphic stela-menhir of schist discovered in 1997 by the Cantonal Service d'Archéologie, Neuchâtel, Switzerland. Length 3.2 m, weight 2800 kg. (Photograph by Thomas Jantscher)

Schists are diverse reflecting their parent and metamorphic mineralogy. Commonly found schists include: quartz-biotite schist, hornblende schist, muscovite schist, chlorite schist and talc schist. Most schists form from shales and the classic metamorphic sequence for this process is: (chlorite) slate → schist, biotite-garnet schist; or staurolite – kyanite schist; or sillimanite schist. Figure 6.28 shows a menhir carved from schist discovered in the Canton of Neuchâtel in 1997 during excavations along the proposed autoroute A5, in the vicinity of the village of Bevaix. The stone is clearly carved as a *stelae* and is dated to the Neolithic Period, ca. 5000 B.P. The length is 3.2 m and the weight is 2800 kg.

The *grade of metamorphism* – low-to-high – is indicated by the mineral assemblage with the micas at the low end and sillimanite at the high end. As we shall see in the last section of this chapter, the firing of clays in ceramic manufacture closely follows the metamorphic mineralogy of shale with the low-high grade minerals – mostly Al_2SiO_5 polymorphs – forming as the firing temperature is increased from earthenwares to porcelains.

6.31.6.2
Thin Section Properties

Gneisses contain layers of quartz and feldspar along with other minerals. In thin sections, the low relief and birefringence of the quartz and feldspar contrasts with the darker minerals. The mineral assemblages of schist are related to the grade of metamorphism. Where the schist has formed from sandstone, the rock will have extensive quartz and feldspar in layers much like the more common micaceous varieties. Because of the deformation of these minerals the grains will be elongated and in the case of the feldspars, lath-like. The PPL and XPL properties will be consistent with those we have listed previously. As the metamorphic grade increases, the XPL of their thin sections produce a riot of highly birefringent colors starting with the micas continuing through the higher temperature minerals of the sillimanite group – staurolite, andalusite, kyanite and sillimanite. Most of these minerals have a relatively high relief and a few have distinctive colors. Staurolite is pleochroic yellow-pale brown. Sillimanite and kyanite have moderate birefringence and are fibrous to prismatic with clear-cut extinction.

6.32
Ceramics and Petrography

A pottery shard may be regarded as a metamorphosed sedimentary rock and clearly amenable to petrographic analysis. Stoltman (1989, 1991) and others (Kamilli and Lamberg-Karlovsky 1979; Kamilli and Steinberg 1985; Rice 1987; Whitbread 1989) in recent years have reiterated the importance of petrography to the study of archaeological ceramics championed earlier by Shepard (1936, 1942, 1954, 1966) as well as by Matson (1960, 1966) and Peacock (1968, 1970). A relatively steady volume of studies, of both Old and New World focus, have continued utilizing tried-and-true geological protocols such as point counting (Chayes 1954, 1956; Galehouse 1971) and optical/instrumental petrographic procedures (Maggetti 1982; Garrett 1986; Freestone and Middleton 1987; Lombard 1987; Vaughn 1990). This chapter has emphasized the optical microscopic identification of rocks and rock-forming minerals as they relate to archaeological geology. In ceramic studies of archaeological interest, these rocks and minerals continue to play an often decisive role in questions regarding ceramic technology, provenance, trade and linkages to belief systems; gender relations and power; as well as art and aesthetics in past cultures. In the next chapter we shall examine instrumental procedures in greater depth while we close this chapter with a continuation of our discussion of optical methods.

6.32.1
Ceramics and "Ceramics"

In the field of archaeological ceramics there is extensive variety in the nature of the specific wares or types of pottery manufactured at any one time in any one place throughout antiquity. To the earth scientist not versed in the complexity of archaeological ceramics or the variety of ways archaeologists look at ceramics, confusion is not a surprising first reaction. In large part, archaeologists rely more on stylistic/morphological variation in ceramic wares than on petrographic parameters. A key concept for archaeological studies, not just for ceramics, although the concept was first used in ceramic analysis is that of "type". A type is a category of archaeological artifacts defined by a consistent clustering of attributes (Thomas 1998b). Variation or consistency in types is used to evaluate the geographic extent and chronological persistence of a particular type, which in turn, is often used as a proxy for cultural variability or persistence. Rarely does the average archaeologist look much beyond the form/shape, color or decoration of a ceramic artifact except, perhaps, to remark on the type of temper used (in earthenwares, mostly). No archaeologist working in Neolithic – Post-Neolithic cultures can ignore the ceramic artifact as: (1) most of these archaeological cultures made and used pottery and (2) by definition, the "Neolithic" is often characterized by the presence of pottery. The earliest pottery appears in the Mesolithic Period in Asia, specifically in the early Jomon Period of Japan, ca. 10,000 B.C. It was independently invented and used in all parts of the world with the exceptions of Australia and, by 5000 B.C., the far north of North America and Eurasia. It is as ubiquitous an artifact in late Holocene archaeological cultures as those of lithics, yet the analyses of these ceramics has focused on the shape and surface rather than on petrological parameters.

This is not to say such archaeological analytical protocols have not led to key insights. One of the first and most famous uses of a systematic examination of ceramic types was that of Sir Flinders Petrie (1904)in his development of ceramic seriation/typological sequence analysis for chronological purposes. Earth science studies of ceramics, specifically petrography provide additional ways or, in the rubric of epistemology, independent evidence, for suppositions and hypotheses about the manufacture and use of ceramics in the past.

6.32.2
Paste and Temper

A ceramic vessel or object – figurine, statue, bead, etc. – is composed of clay and its impurities, generally accidental or accessory minerals and in the case of low-fired earthenwares some form of tempering agent which is often mineralogical in nature. There are a few commonly used terms to denote paste and temper – matrix, fabric, body are frequently encountered. Stoltman

(1991) considers paste to be "The aggregate of natural materials, i.e., clays and larger mineral inclusions, to which temper is later added. . . .". Kamilli and Steinberg's definition where paste "consists of a coarse mineral fraction, a fine-grained matrix, and perhaps some plant material" (Kamilli and Steinberg 1985). Body, as defined by Stoltman (1991) is the "bulk composition of a ceramic vessel including clays, larger natural mineral inclusion in the silt, sand and gravel size ranges, and temper (emphasis added)". Body is equivalent to the term fabric (Shephard 1936, 1954). It is *temper* that discriminates body/fabric from paste (Figs. 6.29, 6.30). Temper, by extension, is that which is not naturally occurring within the paste of a ceramic, e.g., it has been added by the potter.

Temper varies widely in early ceramics from organic materials such as plant material, bone and shell; mineral temper such as rock of many varieties; sand or volcanic ash or tephra; to grog or ceramic fragments reused to temper subsequent vessels. Because temper, other than that of organic origin, is similar to naturally occurring minerals and rock, some authors simply use size as a discriminating factor (Ferring and Pettula 1987; Lombard 1987). The identification of temper, because of its similarity to the materials of the paste, is often-times difficult (Stoltman 1991), but the petrographer must make the attempt wherever possible. Where the distinction is straightforward, the temper becomes a useful tool in distinguishing ceramic types, even in cases such as where the coarser added material may be differentiated from a fine-grained matrix by textural analyses.

In petrographic and textural studies of ceramics four major elements can be readily evaluated:

1. Minerals – major and accessory minerals can be examined by both petrography and particle size studies.

Fig. 6.29. Pottery with micaceous inclusions (light spots) in the paste. (Photograph by author)

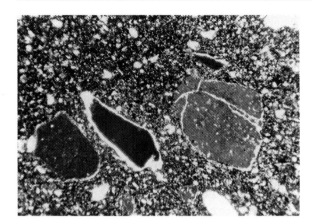

Fig. 6.30. Pottery – thin section view of Late Bronze Age Cypriotic ware showing clay fragments, so-called grog, with obvious shrinkage rims (*light colored*) surrounding them. Cracking observed in the largest fragment is diagnostic of this type of tempering agent. (Courtesy of Dr. Sarah J. Vaughn)

2. Clay minerals – well-fired pottery does not contain true clay minerals since they have reacted and transformed to other minerals.
3. Nonmineral inclusions – both petrography and textural analyses are productive methods.
4. Grain size – by definition, the use of particle size analysis is the use of petrography with the low-power study of grain morphology.

Because clay mineral identification is not typically done with the petrographic microscope we will hold our discussion until the following chapter where more appropriate instrumental techniques are examined. It is important to note that the response of clay minerals and inclusions to firing temperature produces progradational mineralization as in metamorphism as well as melting and recrystallization of the clays and accessories) (Herz and Garrison 1998). A common reaction seen in highly fired wares is the formation of mullite ($3Al_2O_3\ 2SiO_3$) and its antecedent minerals such as kyanite, sillimanite (cf. preceding section). Other minerals such as hedenbergite and fayalite (pyroxene and olivine respectively) form as well, given the appropriate clay matrix. All of these crystal minerals are readily identified by their optical properties – kyanite, in particular, with its rich blue birefringence as well as the bright greens of the pyroxene.

In certain ceramics such as faience and porcelains, the effects of firing are defining elements in the identification of these varieties. From the perspective of optical petrography the procedure is that for the study of archaeological rocks and minerals:

1. Prepare a thin section.
2. Stain, etch or otherwise pre-treat the thin section.

3. Examine the thin section under the low-power PPL/XPL microscope, determining minerals/phases by optical properties and the proportion of each observed.
4. Quantitatively evaluate the thin section using standard point-counting procedures.
5. Examine the thin section under high-power PPL/XPL for relief, color/bire-fringence, etc.

7 Instrumental Analytical Techniques for Archaeological Geology

7.1
Introduction

A simple search of either modern scholarly or popular scientific literature confirms the relevance of chemical analyses to archaeology. Chemical analyses are not only used to determine the chemical makeup of an artifact, but also to reveal clues to age, diet and health when applied to both organic and inorganic materials. Chemistry is equally important in the preservation and restoration of archaeological materials. The first use of chemistry in the study of artifacts can be traced to the great German chemist Martin Heinrich Klaproth's 1796 studies of Greek and Roman coins, other metal artifacts and some Roman glass in which he pioneered the use of gravimetric analysis (Goffer 1980). Gravimetric analysis is the determination of an element through the measurement of the weight of the insoluble product of a reaction with that element. Since Klaproth's first foray there have been numerous examples of the application of analytical chemistry to archaeological materials of all sorts by historical figures in both chemistry and physics – Sir Humphrey Davy, Jon Davy (Sir Humphrey's brother), Jöns Jakob Berzelius, Jon Voelker, Michael Faraday, Marcelin Berthelot and Friedrich August von Kekulè (various in Goffer 1980; Meschel 1980; Pollard and Heron 1996).

According to Robinson (1995), one important function of analytical chemistry is to provide data to other scientists. When applied to materials of art and archaeology the methods must oft times be nondestructive or at the very least, destructive of only a small portion of the material under examination. To Sir Humphrey Davy's chagrin, his 1818 attempt to unroll Pompeiian papyrus scrolls met with less than success when all of the eleven scrolls were effectively destroyed before any deciphering could be done (Herz and Garrison 1998). In contrast, analytical chemistry's exposure of science's most famous fraud – Piltdown – using the quantitative determination of the elements fluorine, nitrogen and uranium in the "fossil" hominid remains and their comparison to both modern and prehistoric bones, in the mid-twentieth century, was the watershed event that brought analytical chemistry into the forefront of subsequent scientific studies of archaeological materials.

The chemical examination of the physical properties of artifacts and other materials of archaeological interest generally requires instrumental assistance. In the last chapter that instrument was the petrographic microscope, which is a very useful tool in the service of archaeological geology. A mineral's optical properties are attributes which aid us in identification and, by extension, the characterization of the archaeological item. With this characterization we can hope to use this mineralogical information to pose pertinent inquiries as Freestone and Middleton (1987) have aptly suggested in the following:

1. The characterization of the material from which an object was made
2. The reconstruction of the technology involved in its manufacture
3. The inference of the place of manufacture or source of raw materials and
4. The changes that have occurred in the object during burial or storage.

The reader can certainly suggest other important questions but these are surely common to most inquiries. Nikischer (1999) has recently reminded us that contrary to popular belief, there aren't any foolproof *single* (author's emphasis) technological innovations that can produce an accurate identification *every* time for *every* mineral (or material for that matter). Frequently two or more confirming techniques are needed.

In the application of analytical techniques to archaeological materials one can seek data at: (1) the molecular level and (2) the elemental/isotopic level. In archaeological geology this would be illustrated by either examining a lithic (marble) sample as to its mineralogical (molecular) form, such as calcite ($CaCO_3$) or assessing the metallurgy of an early Bronze Age axe wherein the presence or absence of the alloying metallic *element* tin (Sn) is critical to its classification, both typologically and chronologically. Likewise, a second important distinction can be made in the analysis of materials – that of (1) qualitative or (2) quantitative. The former simply answers the question, posed in our axe example, "what kind"? A quantitative study asks a second level of question – "how much"? Returning to the axe, if all we want to know is whether the artifact is a copper axe or a later bronze type, then a qualitative study is appropriate to the question. However, if we desire to classify the axe, assuming our analysis determines it is bronze, as to its appropriate typology then we must do a quantitative assessment.

7.2
Analytical Techniques and Their Pluses (and Minuses) for Archaeology

Techniques commonly encountered in the study of archaeological materials include:

1. X-ray diffraction spectroscopy (XRD)
2. X-ray fluorescence spectroscopy (XRF)

3. Electron microprobe/analytical scanning electron microscopy (EMP/SEM)
4. Atomic absorption spectroscopy (AAS)
5. Inductively coupled plasma emission spectroscopy/mass spectroscopy (ICP-ES/MS)
6. Mass spectroscopy (MS)
7. Neutron activation analysis (NAA)
8. Infrared spectroscopy (IR/FTIR)

European laboratories commonly use emission spectroscopy, atomic absorption spectroscopy, X-ray fluorescence and diffraction spectroscopy, and neutron activation . In addition, thermal ion and other types of mass spectroscopy are commonly found. North, and to a somewhat lesser degree, South American laboratories mimic that of the Old World in the diversity of analytical methods regularly used on archaeological materials as do Asian labs, particularly in Japan and China. A recent example of the increasing availability of geochemical facilities to archaeologists and other interests is that of the new EU Geochemical Facility at Bristol University (UK). At the four Bristol labs a comprehensive suite of modern analytical instruments are available to member European Union states together with 15 non-EU members in eastern Europe and the Near East. The instrumentation includes: electron microprobe, ICP-MS, laser ablation ICP-MS, XRF, ICP-AES, XRF, Mössbauer spectroscopy, FTIR, NMR and SEM with analytical energy dispersive spectroscopy (EDS). Surface analysis can be done using Auger Electron, Secondary Ion mass and X-ray photoelectron spectrometers.

From the perspectives of: (1) molecular vs. elemental and (2) qualitative vs. quantitative, each technique has strengths and weaknesses. The X-ray techniques of XRD and XRF are good examples of this. XRD is a powerful instrumental technique in the qualitative and quantitative determination of the molecular nature of a material. By contrast, XRF is a routine method for the qualitative study of elements with atomic numbers higher than $Z = 10$, including both metals and nonmetals. It is also a very sensitive and reliable method for the quantitative study of elements of high atomic weights.

The electron microprobe/scanning electron microscopic instruments (EMP/SEM), like that of XRF, have the capabilities of performing both molecular and elemental analyses as well as those that are qualitative and/or quantitative. In terms of surface analysis, the SEM techniques are highly sensitive to the number of atoms detected, but not their respective percentages (Robinson 1995). Electron spectroscopy is also used only in a semi-quantitative way. Newer electron systems such as analytical SEM models have built-in X-ray energy dispersive spectrometer (EDS) capabilities that allow them to perform qualitative analyses.

AAS or atomic absorption spectroscopy is not used in qualitative analyses except in single element studies. It is, perhaps, the most accurate and sensitive method for the quantitative study of metals with detection limits down to 10^{-12} g. It is not generally used in the detection of nonmetals. Newer AAS instruments have had to expand to rapid multi-element solid sample analyses because of the introduction of inductively coupled plasma spectroscopy (ICP).

In its original form or in newer hybrid configurations with mass spectrometers and lasers, ICP has rapidly challenged AAS with regard to the analyzable range of elements as well as to quantitative sensitivity. Unlike AAS, ICP is restricted to the elemental analysis of materials. By itself ICP is good for the analysis, like AAS, for metals. Coupled with mass spectroscopy, ICP is routinely used for analyses of both metals and nonmetals. It has a wide linear sensitivity range from parts per billion (ppb) to percent levels.

Mass spectroscopy is used in both qualitative and quantitative studies of molecular, elemental/ionic and isotopic forms up to a very high atomic mass. At the quantitative level the mass spectrometer is a commonly used tool for the determination of liquid and gas samples and when combined with lasers, solid samples as well.

Neutron activation analysis (NAA) requires either nuclear reactors or accelerators. The former are common in most developed countries, but the reactors must be lower power, research types rather than the more frequently found power generation reactors. NAA is an extremely sensitive analytical method that can reach the part per billion range (ppb = ng/g) in many materials with ease. It is used for both qualitative and quantitative elemental analyses.

Infrared spectroscopy has been a stable chemical technique for the qualitative and quantitative study of organic compounds. It was the first technique that the author first became acquainted with in its use on the adsorption of contaminants on surfaces. There is abundant literature on this particular technique, but its use in archaeological analyses has not been as extensive as many of the other techniques we will survey. Newer variations on IR spectrometers such as the Fourier Transform (FTIR) designs hold significant promise for the study of archaeological materials, particularly in field settings.

In addition to these methods we shall briefly examine other analytical procedures, such as those listed for the EU facility, whose usage have produced insights into archaeological questions. Electron spin resonance (ESR), cathodoluminescence (CL), magnetic susceptibility will be mentioned relative to specific archaeological findings such as provenance studies of marble, cave sediment and climate, to specifying materials used in the Dead Sea Scrolls. A major consideration in discussing this particular suite of instrumental techniques is their common usage in the analysis of materials of archaeological interest. All have long track records in both their theoretical underpinnings and their practical usage. For the archaeological researcher other considerations enter into the selection of the instrumental technique or techniques for his/her particular needs. Key among these practical considerations are:

1. nondestructiveness of the technique
2. accuracy and reliability of the results

SELECTION CRITERIA

3. availability of the technique $\}$ SELECTION
4. cost. $\}$ CRITERIA

The first consideration – nondestructiveness, by the technique, of the archaeological sample of interest – is not a trivial one. Because of the rarity and
uniqueness of many archaeological objects, any damage to them beyond that
of a natural diagenetic origin while unexcavated, is generally to be avoided.
Archaeologists and museum curators are loath to expose unique collections to scientific analyses that pose the threat of damage or destruction
to the artifacts. As an archaeologist the author has had to deal with just this
problem when presented with the choice between preservation of an excavated item and the determination of a much-needed absolute date. In this specific case the decision to sacrifice the archaeological sample – a prehistoric
corn cob, rare in the context in which it was found – proved too compelling
when its preservation was weighed against the need for a reliable age determination of an incidence of prehistoric farming in the American southeast.
This is not usually the case in other situations. Preservation of the analyzed
archaeological material is considered paramount.

The accuracy and reliability of the analytical results are key considerations
in the choice of an instrumental technique. All the techniques we shall discuss
have varying levels for the detection of elemental or mineralogical parameters. Some can regularly achieve accurate determination of trace levels (less
than or equal to 1 µg) of interest. Others can only achieve accurate results at
the percentage level of analysis. It is perhaps important to digress here and
eliminate any confusion about what is meant by "accuracy" and its oft-
confused adjunct "precision." Accuracy refers to the *correct* measure of a
particular physical property such as temperature, mass, age, etc. The precision of that measurement refers to its replication with reliability and most
important repeatability. If the true mass of an artifact is 30 g, then a measurement of 20 g is obviously in error. Should an instrumental technique
repeatedly measure the instrumental mass as 20 g to three or four place
precision, it is still an inaccurate measurement. However, it is a "precise" inaccurate measurement.

Therefore, reliability is measurement-bound in that one seeks to achieve
evaluations with instrumental methods. The seven listed techniques have
established reliability to differing measurement levels. Of them, neutron activation analysis (NAA), a gamma-ray spectroscopic technique can lay claim
to the crown for the most accuracy and precision at trace element levels of
over 70 elements. ICP, inductively coupled plasma spectroscopy, is rapidly
approaching NAA's analytical standard and will likely supplant the older
technique by virtue of greater availability and less cost per analysis.

The last two parameters – availability and cost – are likewise, important to
the choice of analytical technique. Again, in comparing NAA and ICP, they are
comparable in terms of measurement accuracy, precision, and repeatability.
The availability of NAA is constrained in part by that of nuclear research

reactors, but more readily available (and less politically controversial) particle accelerators can provide adequate neutron fluxes for most archaeological studies. Nonetheless, both these types of NAA facilities are capital-intensive and limited by real concerns about radioactive wastes produced by their use. As ICP and other high precision techniques produce little contaminates, like radionuclides, and have the advantage of lower costs and size, they will likely supplant NAA as an analytical technique of choice in the near future.

7.3
X-Ray Diffraction

Spectroscopy is the determination of the physical properties of a material – rock, mineral, biological, solid, liquid, or gas phase – by the absorption and/or emission of energy. The spectra of absorption and emission can be characteristic of a material, thus allowing its identification and measurement. The following sections discuss the properties and characteristics of individual instrumental techniques beginning with the X-ray spectroscopic techniques of XRD and XRF.

A colleague has called XRD the "first line of defense for archaeological (analytical) analyses" (Paul Schroeder, pers. comm.). XRD is a technique that yields a mineralogical identification rather than an elemental characterization. XRD is a monoenergetic spectroscopic technique based on Bragg's law:

$$n\lambda = 2d \sin \varphi$$

where n is the order of wavelengths, λ the wavelength, d the interatomic spacing in a crystal and φ the diffraction angle of the X-ray leaving the mineral's lattice (Robinson 1995). This is shown in Fig. 7.1. For XRD to work, the archaeological material *must* be crystalline. With the exception of glassy materials like obsidian, most minerals are crystalline.

In XRD analyses, monoenergetic collimated X-rays are allowed to fall on specimens and a proportion will be diffracted at angles (φ) depending on the crystal structure of the specimen (Fig. 7.1). An essential feature for diffraction is that the distance between the scattering points is the same magnitude of the wavelength of the incident radiation. XRD works only on crystalline materials, as we have said, as there is no refraction in a glass because there is only random inter-crystalline structure. The orderly arrangement of atoms in crystalline minerals is seen in the repeating structure of the unit cell. Every mineral has a unit cell dimension that is specific to it. In practice, the XRD device scans a sample, generally a powder, with a fixed X-ray wavelength of 1.54059 Å, the $K\alpha$ X-ray (L shell to K shell) for copper. By comparison the unit cell spacing for silica is 1.6 Å.

XRD has a long and successful record in mineralogical identification, most notably the clays and phyllosilicate minerals – smectite (montmorillonite), illite, halloysite, vermiculite, chlorite, and serpentine. Because of the impor-

11.12

Fig. 7.1. X-ray diffraction.
Above Diagrammatic view of
an XRD spectrometer, *below*
illustration of the XRD
principle – Bragg's law

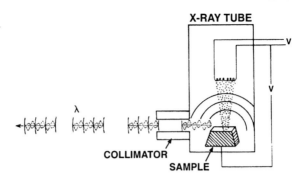

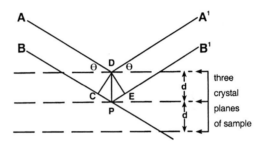

Diffraction in a crystal according to Bragg's law, $n\lambda = 2d \sin \Theta$.

Great for clay ID in ceramics

tance of ceramics in archaeology, the utility of a spectroscopic technique which identifies and quantifies clay varieties should be readily apparent. In XRD the angle φ, of the diffracted X-ray radiation is different for the different crystals due to the various interatomic distances. The diffracted radiation is measured as shown in Fig. 7.2a,b. The detector is swept through 2φ to get our measurement of φ which can yield very good resolution (μg/g-range). XRD is commonly used to study fine-grained minerals too small to be examined optically.

For metallurgical studies, XRD has the capacity to evaluate the state of an object's anneal. In well-annealed metals there is clear crystal order yielding sharp diffraction lines. If the object has been hammered, drilled or otherwise worked as is typical of early metallurgy, then this will contribute to the blurring of the diffraction spectrum due to the effects on crystal order. In addition, corrosion can be examined in the artifact. Recently, XRD, coupled with XRF-EDS, has been used to analyze pigments from Trajan's Column (Del Monte et al. 1998). In this study, the researchers isolated hematite and *minium* (HgS) or cinnabar that had been applied over an underlayer of plaster indicating the ancients' use of color for this monument. Pretola (2001) has demonstrated XRD's potential for determining varieties of cherts and their possible

✳ anneal ∴ TO HEAT TO REMOVE OR PREVENT INTERNAL STRESS.
• TO TOUGHEN OR TEMPER

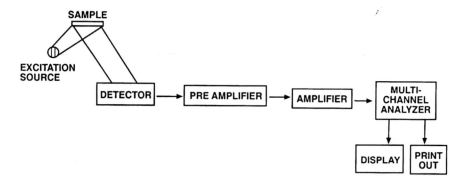

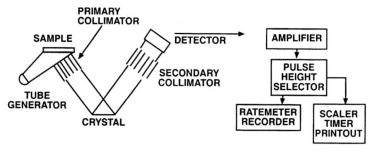

Fig. 7.2. Diagrammatic representation of an energy-dispersive spectrometer (*above*), and a wavelength dispersive spectrometer (*below*)

source locations. In this study XRD's ability to differentiate variations in the mineralogy of silicates, in particular "moganite", a quartz *polymorph* much like those of calcium carbonate, e.g., calcite and aragonite (moganite is not currently approved as a formal mineral name Pretola 2001). Since this quartz variety cannot be easily discerned optically, XRD is the instrumental method of choice in its study. An interesting aspect of this quartz form is its diagenetic alteration to quartz over time ($>10^6$ a), implying that cherts older than 100 Ma will have little or no amount of the polymorph (Heaney 1995). This *metastability* of a mineral is similar to that of anhydrite/gypsum and calcite/aragonite (cf. Chap. 6). Jercher et al. (1998) have used "Rietveld" XRD analysis for the study of Australian aboriginal ochres. The Rietveld method is a powerful method of fitting observed spectra to structural models of specific mineral phases (Rietveld 1969). Although it is a powerful technique, XRD cannot differentiate certain minerals from others such as gold and silver. In addition, the presence of certain elements and their oxides, such as those of iron, can cause nondirectional scattering and increase background interference or scattering. Alumina and opaline silica can cause these effects as well. Any significant increase in background scattering decreases the sensitivity of the technique for other minerals. Particle size is critical as well. Particles greater than 10 mm in diameter create a reduction in diffraction intensity as do smaller particles (0.02 mm). Heterogeneous samples must be treated carefully because of these size effects. Some workers use internal standards to offset nonrandom scattering, self-absorption and other problems. One can then examine diffraction-to-intensity ratios for the sample and the standard to obtain mineral composition by weight (Whittig and Allardice 1986). The archaeological geologist must be aware of these effects and the realities of XRD analysis when deciding to proceed with this line of instrumental inquiry.

Preparation procedures of samples for XRD analyses are determined by the nature of the material to be studied. In archaeology the material will generally be either ceramic or a lithic material such as flint or marble. If the material is a sediment or soil, preparation will be much the same as for a ceramic particularly if the pottery is earthenware. Finer-grained wares, including porcelain, will be less sensitive to particle size considerations. Separation of textual classes from ceramics will produce relatively distinct fractions of the paste or body and again in the case of coarser earthenwares, the temper. Specific techniques for particle size determination were discussed in Chapter 5. The reason samples are separated as textural sizes is that almost all XRD studies samples are mounted as either (1) a random powder or (2) an oriented aggregate. The powder mount, favored by most labs, enables one to obtain as many possible diffraction spacings in the minerals present. Relative diffraction intensities from a random powder mount more nearly proportional to the actual amount of crystals present than those obtained in an oriented aggregate.

7.4
X-Ray Fluorescence XRF

X-RAY BOMBARDMENT
WITH EMISSION SPECTRA

In X-ray fluorescence spectroscopy (XRF), the specimen is bombarded by X-rays from a conventional X-ray tube and emits a secondary (fluorescent) spectrum of X-rays whose wavelengths are *characteristic* of the elements present in the specimen.

It is important to enlarge on the distinction between the wavelength dispersive spectroscopy (WDS) and energy dispersive spectroscopy (EDS). WDS is the older variety of XRF wherein the characteristic wavelength of the secondary X-rays produced by the X-ray rather than by electrons (primary excitation) is measured. Wavelength dispersive spectrometers are better suited to accurate quantitative analyses. By use of a diffracting analyzer crystal in an accurate circular geometry (Roland's circle), coupled with exit slits (on the crystal), only X-rays of a specific wavelength are counted on a proportional counter. Bragg's Law can be used as with XRD (Goffer 1980; Nikischer 1999).

Energy dispersive spectroscopy (EDS) is the measurement of the secondary X-ray's energy that is characteristic of a given element by use of a scintillation detector, either composed of silicium/lithium (Si/Li) or germanium/lithium(Ge-Li). These crystals will fluoresce when struck by secondary X-rays generating a signal which is amplified and sent to a multi-channel analyzer. In general, EDS has a 0.5% detection limit making it a fast but qualitative technique for elements with Z > 5. This detector system experiences a lot of "dead time" and does not work well for the heavy elements because their X-rays typically have high energies which allow them to pass through the detector. EDS is a good method for the rapid, qualitative characterization of a sample to determine the variety of elements present. The SEM (scanning electron microscope) is optimized to use the EDS spectrum for image formation.

XRF can perform in both modes and it has been in WDS that most studies prior to the 1990s have been made (Jones 1991). Due to advances in both detector crystals and analytical software, EDS has seen a significant increase in its usage because it is: (1) faster than WDS, (2) has detection limits of 0.1% by weight and (3) resolves more elements with fewer peak overlaps than WDS (Nikischer 1999). WDS has better detection limits, approaching the parts per million (ppm = µg/g) range, but significant time is required compared to EDS? (Potts et al. 1985; Nikischer 1999). In many modern XRF instrument both capabilities are available. As an EDS system, XRF can be applied to both powder-like and solid samples. In qualitative EDS studies very little sample preparation is required such as those used in the aforementioned study of Trajan's Column (Potts et al. 1985; Del Monte et al. 1998; Nikischer 1999). Rock can be placed in the XRF sample holder, up to reasonable sizes; soils and sediments simply poured into Teflon cups and placed in the spectrometer; and ground ceramic material can be spread on Mylar tape and scanned by the XRF. Some commercial laboratories grind and press the powder into disks similar to XRD studies. Others will fuse the sample into a glass, with a borate flux, that forms a disk at 950 °C (Jones 1991).

The detection of several elements (over a dozen) is readily done by XRF. These include Mn, Fe, Ni, Cu, Zn, As, Br, Rb, Sr, Zr and Pb. Sodium and magnesium are not easily done and elements below Z = 20 are somewhat more difficult than those just noted. Figure 7.3 shows the relationship between detection and atomic number. XRF is extremely useful for the analysis of both metal and nonmetals with Z > 12 and has the distinct advantage over both emission spectroscopy and AAS because neither method can directly determine nonmetals.

Scott (2001) has demonstrated the use of XRF in scanning both ceramics and paintings at the Getty Museum. This scanning X-ray microfluorescence system was used on an Attic lekthyos (vase) and a seventeenth century Dutch painting measuring five elements on the former and seven elements on the

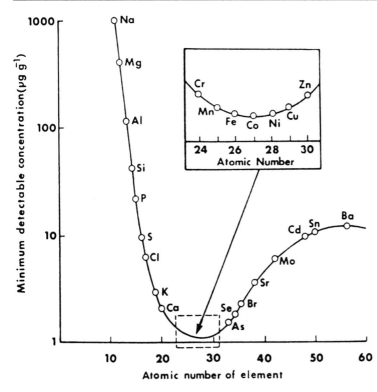

Fig. 7.3. A guide to minimum concentrations of elements detectable using X-ray fluorescence (XRF)

latter. In particular, this system can examine lead white and vermilion along with the high Z elements. In the instances where SEM-EDS has problems with peak overlap such as with Ca Kα (4.04 KeV) and Sb Lβ (4.38 KeV), the XRF can rapidly measure the calcium, a common mineral pigment of "bone black" (Scott 2001). For rapid assessment of painted surfaces XRF, UV and IR together with XRD form an excellent instrumental ensemble. Guineau and coworkers have examined black pigments in paints from prehistoric cave paintings at Pêche Merle in France (2001). XRD analysis of the paint pigments indicated the prehistoric artist(s) used the manganese minerals hollandite ($BaMn_8O_{16}$) and romanechite ($BaMn_9O_{16}(OH)^4$) mixed with common black.

XRF-EDS was successfully used to demonstrate the metallurgy of the copper axe (Fig. 7.4) found with the Iceman mummy in 1991. In the first days of this important archaeological discovery, (Spindler 1994), the distinction between a Copper Age and Bronze Age axe is significant, representing in the European area a difference of a millennia or more. After the excitement of the discovery subsided, Spindler submitted the axe to an XRF analysis at the University of Innsbruck. The result was quick and decisive – no tin was found.

Fig. 7.4. Replica of a Chalcolithic axe shown with the mineral

The axe was copper and the Iceman's artifact was one of the oldest such items ever found in the late Neolithic Period, ca. 3500 B.C.

7.5
Electron Microprobe Analysis and *EMP.*
Proton-Induced X-Ray Emission *PIXE*

The combination of thin-section petrography with instrumentally based chemical analysis has proven to be one of the most useful associations for archaeology in recent years (Freestone and Middleton 1987). As we have seen in Chapter 6, thin-section petrography of archaeological materials can provide significant insights. The use of electron microprobe, proton-induced excitation emission spectroscopy and the analytical (SEM) scanning electron microscope can enhance and extend the information obtained by optical petrography.

All the systems – EMP, PIXE and SEM – are "scanning" systems which implies the particle beam, electrons in EMP and SEM, protons (hydrogen ions) in PIXE, can be focused and rastered (X-Y) across the surface of an example. This use of the electron/proton as the exciting particle is the difference between these instruments and the X-ray systems. In addition, the interaction of the incident beam, while physically similar to XRF, has the advantage of deeper and more energetic penetration of the subsurface of the material under investigation. The energy and mass difference between XRF's photons, SEM/EMP electrons and PIXE's protons means that EMP/SEM can only excite the near surface atomic layers, due to the dispersive nature of the incident beam; XRF significantly deeper (>5 μm) by virtue of its high energy, collumi-nated X-ray beam and PIXE the deepest (>15 μm) of all. Research at the Ruhr-

Universität on Roman coins using both XRF-WDS and EMP-EDS illustrates this. The surface analysis of these coins typically found a penetration depth of 3 µm using the latter technique while a penetration of 30 µm was achieved using XRF-WDS (Klockenkämper et al. 1999).

Analysis by EMP and PIXE has become widely used in elemental analysis of both geological and archaeological materials (Reed 1996; Henderson 2000). In electron microprobe analysis (EMPA), electrons from a filament are accelerated to about 30 KeV, directed at a specimen and focused to a spot 0.001 mm in diameter. When the beam strikes the specimen, X-rays are produced that are characteristic of the elements present in the specimen. These X-rays are analyzed by a spectrometer as in the XRF technique. In PIXE analysis the electron is replaced with a proton as the incident particle. The technique is based on the excitation of the surface atomic electron shells by the particle beam.

In theory, both EMP and PIXE are nondestructive techniques, but this is only relative to the sample size. To acquire the specimen or sample may not be nondestructive. EMP and PIXE samples are small, less than a few millimeters as a rule, in order to fit in the sample chambers of the analytical devices. EMP sample chambers and early PIXE models are kept under a vacuum so the particles are not scattered or absorbed by air particles in their paths. As a result small samples are taken from the larger object resulting in some damage to the object or artifact. Newer PIXE designs allow the particle beam to be operated in air allowing for analyses to be performed without recourse to destruction of the object.

An advantage of either technique is their capability of analyzing small areas in great detail and with precision. Both EMP and PIXE can achieve lower levels of elemental detection than XRF – typically ppm levels for the former and percent levels for the latter – thus achieving good accuracy in the quantitative results. Using the EMP, researchers have the advantage of (1) quick, quantitative elemental determinations, (2) use of the dispersive nature of X-rays for a semi-quantitative characterization of large suites of elements and (3) the production of secondary, backscattered electron (BSE) images of the sample surface. The back-scattered (secondary) electrons (BSE) are the result of elastic scattering of the primary electrons in the EMP. BSE pass through the atomic electron cloud and change direction without appreciable energy loss. They may travel into the sample or escape from the surface in fluxes that are proportional to the atomic number (Z) of the material. The higher the atomic number, the smaller the mean free path between scattering events and the higher probability for scattering (Rigler and Longo 1994). As a result, lower Z elements in a material allow primary electrons to penetrate deeply and reduce the number of BSE. Conversely, higher Z elements retard the beam's electrons and emit larger numbers of BSE. This difference in BSE emission as a function of atomic number can, and is used, to map the distribution of different elements across the surface. Images produced by BSE have excellent resolution and provide the SEM/EMP observer with surface compositional detail by rastering, in an X-Y pattern, the beam over the sample.

The analytical SEM combines the superior imaging capability of electron microscopy with semi-quantitative chemical analysis using EDS. In ceramics and sediments/soils the SEM can examine the microstructure of these materials-grain morphology, crystal size and shape, etc. Coupled with this imaging capability the analytical SEM's EDS can move, seamlessly, to a semi-quantitative analysis of crystalline and glassy forms as minerals and slags (Freestone and Middleton 1987). Studies of Bronze Age coinage corrosion products, by SEM-EDS, have elucidated mechanisms for the formation of two differing patinations (Robbiola and Fiaud 1992). The principal differences in the analytical SEM and the EMP are in accelerating current voltage, higher in the EMP, and the beam diameter, a larger diameter in the EMP.

The source of electrons used by the SEM or EMP is a heated filament just as with the XRD and XRF devices. These electrons are not allowed to strike a metal plate to produce X-rays, but are accelerated by an electron gun (Fig. 7.5a) much like a television set and in the case of SEM, are rastered by the gun across the sample. The system is under vacuum. For imaging purposes, the BSE spectrum is used (Fig. 7.5b). The relationship between penetration depth, energy and the sample's atomic number (Z; Fig. 7.5c) is important. The higher the atomic number, the smaller the mean free path between electron scattering events (e.g., 528 Å for Al vs. 50 Å for Sn at an accelerating voltage of 30 KeV) and the higher the probability for scattering. Normal voltages for EMPs are 10–15 KeV, so this example is quite high by comparison. New SEM models can have voltage ranges of 40–200 KeV. As noted earlier, the difference between the SEM and the EMP is that the former is optimized to use the EDS spectrum for good images while the EMP is optimized to provide analytical results, e.g., an EMP beam is focused at 0.1–0.5 μm while the SEM focuses at 10–1000 nm!

Polished thin sections, such as are used in optical petrography (cf. Chap. 6), are generally used with the EMP, thereby achieving greater precision and accuracy – true quantitative analysis. An SEM optimized for obtaining images, even with the EDS detector present, and without the use of polished thin-sections produces significant surface spectral interference reducing the accuracy and precision of the data.

PIXE instruments use neither photons (XRF) nor electrons (EMP, SEM), but rather more massive protons (2000× the electron mass). Accelerating these particles to high voltages allows greater subsurface penetration with very high analytical resolution (ppb range). All these techniques use the characteristic X-rays for these analytical procedures (Fig. 7.5b). Besides the X-rays, a major component of the energy emitted, by a sample, as a result of primary beam interaction with the material are secondary electrons. Use of a PIXE microprobe is illustrated in a study of Pre-Columbian metallurgy by Palmer et al. (1998) wherein quantitative analyses of minor (>100 μg/g or 0.01%) and trace (<100 μg/g) elements were done on copper bells or "crotals". Combined with

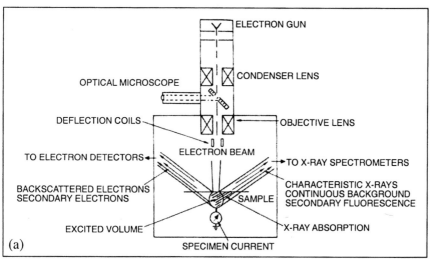

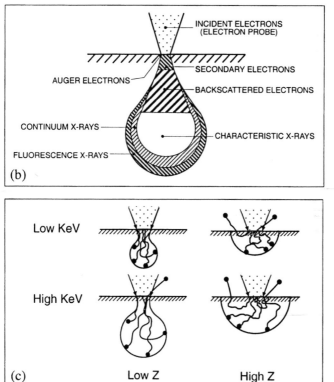

Fig. 7.5. Schematic diagrams of (a) the components and emissions of an electron microprobe, (b) close-up of the excited surface volume of a sample bombarded by the electron beam, (c) the relationship of electron energy (E) and the atomic weight (Z) (cf. Fig. 7.3) in the excited volume

analytical SEM results from the same artifacts, the study found six compositional types within the 24 bells inferring several prehistoric workshops of a trade-linked Hohokam–Anasazi period in the American southwest.

7.6
Atomic Absorption, Inductively Coupled Plasma/ Atomic Emission Spectroscopy (AES)

Atomic emission spectrometry (AES) relies on the property of the spectral emission of photons specific to an element at a rate and proportion to the amount of that element and the temperatures of the flame of a burner/furnace similar to that of our AAS example (Fig. 7.6). In AES types of analysis the intensity of the selected wavelength (λ) is compared to that of some standard. As anyone who has done the simple flame emission test for sodium knows, the color (yellow) of the flame is directly a result of the wavelength of photons emitted by the stimulated sodium atoms. Flame temperature is central to this stimulated emission. In most AES systems the physics require increased (flame) energy as the wavelength (of the element) decreases. Shorter wavelengths always correlate with higher energy – X-rays are longer in wavelength than gamma rays, for example.

Atomic spectroscopy is the most widely used technique for the determination of elements in most materials, especially the metallic elements. The most common techniques include (1) flame atomic absorption using the hollow cathode lamp (HCL), (2) graphite furnace AAS and (3) inductively coupled plasma/atomic emission spectroscopy. Coupled with mass spectrometry, ICP has become one of the most versatile of the modern atomic spectroscopic techniques becoming both the atomizing and ionizing device in these hybrid systems.

The HCL or flame AAS has been the standard unit of small labs. It is an efficient and robust instrument for the quantitative study of a single element at a time to relatively acceptable levels for most metallic elements of interest (~μg/g). A drawback of the flame method is difficulty in determining refractory elements like boron, tungsten, zirconium, or tantalum because the flame is not hot enough. The graphite furnace method typically has an advantage

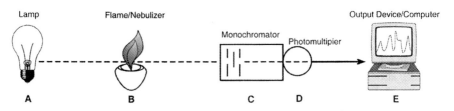

Lamp   Flame/Nebulizer   Output Device/Computer

Monochromator
Photomultipier

A   B   C   D   E

Fig. 7.6. Schematic diagram of an atomic absorption spectrometer (AAS)

11.17

of two orders of magnitude in detection limits, but suffers the same problem with refractory elements as with the flame method.

AAS has a long record in archaeological studies far too numerous to recount here, but by way of illustration a recent study of proto-historic/historic Wichita ceramics in eastern Oklahoma demonstrates the utility of AAS. Singleton et al. (1994) examined 8 elements in the 62 sherd samples from a large American Indian site to examine the heterogeneity in the clays and the linkage of this variation (or lack of same) to manufacture/trade practices among the members of this regional tribal group. The use of AAS was made on an economic, as well as a methodological, basis – to wit less expensive than say, NAA, and certainly more accessible in terms of the ubiquity of AAS instruments in most college chemistry units. True to its long record of success in archaeological chemistry, this use of AAS in concert with modern multivariate statistical analysis discriminated three different locations of manufacture that could reasonably be correlated with cultural behavior of the groups responsible for the ceramic assemblages.

To achieve the temperatures necessary for viewing elements with analytical wavelengths below 250 nm, spectrometers are beyond the capacity of chemical flames such as those of air/nitrous oxide and acetylene (Ure 1991). To achieve these temperatures, radio-frequency plasma sources have been developed where the plasma flame (Fig. 7.7a) is produced (induced) by a high-energy radio source, e.g., inductively coupled plasma or ICP. With flame temperatures of 4000–10,000 K, emission spectra of elements lying below 200 nm are achieved.

LeBlanc (2001) reviews current solid-sample methods approved by the US Environmental Protection Agency (EPA) that involve acid digestion-extraction for 30 analytes from solid phase samples. He discusses five so-called "dilute-and-shoot" methods of sample preparation for AA or ES analytical systems. These are:

1. Soxhlet extraction (EPA 3540C) – used for over 100 years, this technique uses a porous cellulose thimble with the solvent below its boiling point to digest the sample over a 16–24-h-period at 4–6 cycles/h.
2. Automated Soxhlet extraction (EPA 3541) – this procedure has a higher cycle time – every 2 h, yielding more samples for analysis.
3. Pressurized fluid extraction (EPA 3545) – sample cell is loaded with solvent and constant heat and pressure applied. The cell is flushed with clean solvent and then purged with nitrogen gas yielding the extract/analyte. This method has the advantage of producing samples at a rate of every 10–20.
4. Microwave extraction (EPA 3546) – the sample is sealed in the extraction vessel with the solvent above its boiling point. It compares in speed to EPA 3545 and has a performance comparable to the Soxhlet method.
5. Ultrasonic extraction (EPA 3550C) – the solvent and sample are exposed to ultrasound for 2–3 min and centrifuged. It does not compare well with

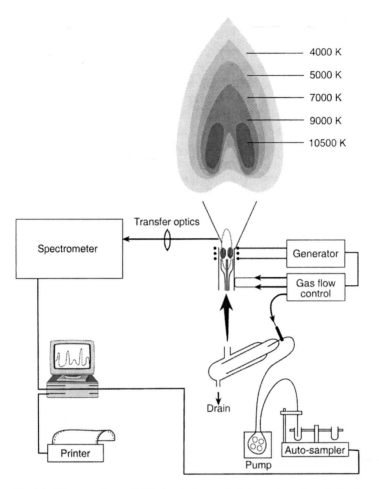

Fig. 7.7. Diagram of a typical inductively coupled plasma/atomic emission spectrometer (ICP-AES) showing approximate temperatures of the plasma flame regions in degrees Kelvin

the Soxhlet method in terms of analyte extraction. Extraction of analytes from paleosol samples is not very thorough.

The vaporized/atomized sample is injected with a carrier gas such as argon and the particle atom experiences ionization as well as excitation. Excited atoms emit photons proportional to the atom's electronic structure and the energy it experiences. Because of the high temperature in the central plasma flame, high ionic-excitation emission occurs. The basic ICP-AES device is shown in Fig. 7.7. Rychner and Kläntschi (1995) have used ICP-AES to good purpose in a comprehensive study of bronzes from the Middle and Late Bronze Ages. In 941 spectrometric analyses, the researchers, using 11 metallic elements, were able to address questions of copper technology, typology

[handwritten: 11.17]

and provenance for these Swiss finds. Linderholm and Lundberg (1994) utilized ICP-AES in their attempt to characterize archaeological soils using 12
elements. ICP is a much faster technique than AAS, being capable of multi-
element analyses, whereas AAS is not. To achieve multi-element analytical
speed while gaining the detection sensitivity of graphite furnace AAS, ICP was
combined with the mass spectrometer. Before discussing the ICP-MS and its
variations (Fig. 7.8) we are best served by first discussing mass spectroscopy.

7.7 Mass Spectroscopy

[handwritten: ASTON (1919)]

[handwritten: Thomson in 1913]

Positive ions in an electrical discharge are repelled from the anode toward the
cathode as noted by the discoverer of "positive rays" Sir J.J. Thomson in 1913.
In 1919, Francis W. Aston invented the principle of the modern mass spec-
trograph where positively charged particles (ions) are placed in a varying
electrical potential and, thus accelerated through the magnetic field (H). The
ions curve according to their respective masses according to the relation:

$$R = mv/eH$$

[handwritten: Radius = ((mass)(velocity)) / ((charge)(mag. field))]

where R is the radius of the curvature; m is the mass of the ion; e the charge,
v is velocity. In gas source instruments, the ions strike targets called Faraday
Cups and their incidence and mass are measured and displayed. Today, software allows a rapid display of the masses and their amounts, such as the
important isotopes of carbon – ^{12}C, ^{13}C and ^{14}C. Isotopes are nuclides with
differing atomic masses of the same element, such as the carbon isotopes
just mentioned. They differ in the number of neutrons present in the nucleus
while the proton or atomic number, Z, remains the same.

Mass spectroscopy is an analytical technique that provides both qualitative
and quantitative data on the elemental and molecular forms of inorganic and
organic materials. If the material can be readily vaporized, then mass spectroscopy can directly detect molecular weights from single digit values to
those of six figures typical of the complex organic varieties. As a quantitative
tool it can evaluate concentrations at the ppm level. The real value of the mass
spectrometer is its ability to resolve particles of differing masses, a property
termed resolution.

The early mass spectrometers and indeed most analytical types are based
on the separation of mass by magnetic fields. The modern Nier-type of mass
spectrometer, developed by Alfred Nier in 1938, consists of three parts – a
source of a positively charged ionic beam, a magnetic analyzer and an ion
collector. By varying the magnetic field or the accelerating voltage one can
generate a mass spectrum consisting of a series of peaks and valleys where
the respective peaks represent an isotope whose abundance is proportional to
the peak height (and area). A common variant of this design is the 60° magnetic sector mass spectrometer used in stable isotopic studies of elements

*[handwritten: MASS SPEC
1.) Source of positively charged ion beam
2.) Magnetic Analyzer
3.) Ion Collector]*

11.17

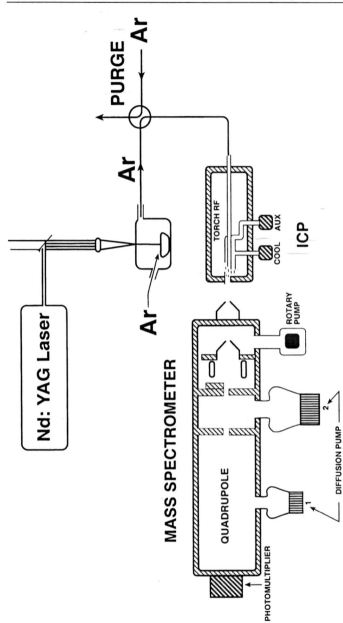

Fig. 7.8. Schematic diagram of a laser-ablation–inductively coupled plasma–mass spectrometer (LA-ICP-MS)

such as hydrogen, carbon, oxygen, nitrogen and sulfur. For heavier masses the double focusing thermal ionization mass spectrometer (TIMS) is used.

These instruments are also termed "gas-source" (carbon/oxygen isotopes) and "solid-source" (strontium) mass spectrometers.

Two nonmagnet instrument designs commonly found in today's labs are (1) the quadrapole mass spectrometer and (2) the time-of-flight mass spectrometer. The quadrapole MS uses four electrodes or poles, hence the name, to produce an oscillating magnetic field wherein the ion, if the mass and the oscillatory frequency are compatible, will pass through to the detector without any deviation. The quadrapole mass spectrometers are smaller than the larger magnetic sector designs and generally cheaper. They do have the resolution of the conventional designs, but not the accuracy or precision. The time-of-flight designs (TOF-MS), likewise do not use large magnets to separate the charged ionic particles. The time-of-flight MS separates the ions by differences in their velocities after they leave the acceleration chamber (Fig. 7.9). This type of mass spectrometer is a configuration used more and more in geochemical analyses – the SIMS or secondary ion mass spectrometer. This instrument uses the velocity of the secondary ions ejected from the surface of a sample, which is indicative of their respective masses, in determining their arrival times at the detector. TOF-SIMS have a large mass detection range. Patel et al. (1998) have used TOF-SIMS in the surface analysis of archaeological obsidian to determine its utility in determining weathering

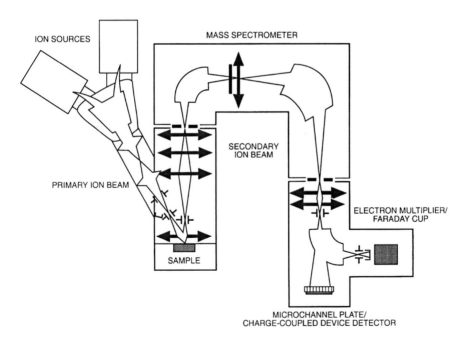

Fig. 7.9. Diagram of a typical modern time-of-flight mass spectrometer

profiles as a substitute for light microscopic measurement used in hydration dating. Perhaps most importantly, the SIMS-MS and TOF-MS units, in concert with high resolution TIMS and SHRIMP (super high resolution ion microprobe) can rapidly assess the isotopic p ratios and amounts of uranium and thorium – a "revolution" (Stos Gale 1995) in uranium series disequilibrium dating.

A new analytical instrument design pairs the Nier-type MS with gas chromatography (GC; Robinson 1995). While not treated in depth here, chromatography analysis is a proven and powerful analytical technique for organic macromolecules of archaeological provenance, such as residues on prehistoric vessels and even those purported to be the blood residues of butchered animals on ancient tools (Loy and Dixon 1998; Malainey et al. 1999). The GC-MS Nier-type systems have the advantage, in resolution, over quadrapole designs and, when coupled with a pyrolysis furnace for sample combustion permit the rapid analysis study of samples to be readily done at well below microgram levels. The GC has been coupled to the Fourier transform infrared spectrometer (FTIR) as well. This device provides a rapid approach to characterizing complex mixtures. An interesting avenue of study in regard to the GC-MS systems is the splitting out of carbon and other light stable isotopes for later analysis in AMS systems. The realization of the last method has been difficult given the need for microgram quantities in AMS machines and the much smaller samples typically produced in GC studies.

Isotopic analysis of the lighter elements – hydrogen, carbon, oxygen and nitrogen – in both rock and organic samples is done using a gas source, magnetic sector mass spectrometer. Because elemental and isotopic differences exist in food chains of both marine and terrestrial organisms, the isotopic analysis of inorganic and organic constituents of bone can provide a record of long-term dietary intake. The isotopes of choice in these types of archaeological studies are carbon and nitrogen. The carbon isotopes of ^{12}C and ^{13}C allow us to evaluate the metabolic paths of the two photosynthetic classes of plants – C3 and C4. The C3 plants include all the trees, shrubs and grasses from temperate and forested environments while C4 plants are grasses from the subtropics and include maize. Important C3 grasses are the wheats and other cereal grains domesticated in antiquity. C3 plants tend to have more reduced (or negative) carbon isotopic values compared to those of the C4 plants, e.g., –26.5‰ vs. –12.5‰ (Van der Merwe 1992). Although significantly fractionated in an organism, these isotopic ratios track in relatively predictable proportions such that an animal or human consuming these plant types will exhibit values that have direct dietary implications.

Workers like Van der Merwe (1992), Price (1989), Sealy et al. (1985) and others have demonstrated that the mass spectroscopic analysis of carbon, nitrogen and strontium isotopic ratios can differentiate the types of foodstuffs being consumed by ancient populations – from both terrestrial and marine resources. Beyond paleodiet, studies of paleoclimate have benefited from the

WHEAT = C3 !

11.17

evaluation of oxygen isotopes in marine foraminifera which act as proxies for sea surface temperature (SST) – the heavier ^{18}O is enriched in colder climates (SST < 20 °C) – by the preferential evaporation of the lighter ^{16}O isotope. Herz (1987) has basically re-invented the field of marble provenance by using carbon and oxygen isotopic ratios in the study of marbles. A recent student of Herz has continued these isotopic studies in a comprehensive study of the Pentelic quarries of Athens (Pike 2000).

7.8
Inductively Coupled Plasma – Mass Spectroscopy

The "marriage" of the ICP to the mass spectrometer has produced one of the most versatile analytical instruments for use on archaeological problems. By combining the ICP's high ionic efficiency, almost 100% for most elements of the periodic table, and low incidence of doubly charged ions (a problem with other conventional MS), with quadrapole and magnetic sector mass spectrometers the mass spectra are relatively simple and elements and their isotopic ratios can be straightforwardly determined. For example, a 20-ms-scan covers the entire mass range of the periodic table with a sensitivity (0.1–0.2 ng/g) comparable to AAS. The mass spectrometer becomes the detector for the ICP. The greatest advantage of the ICP-MS is that it can be used to analyze both metals and nonmetals. Kennett et al. (2001) have termed the increasing use of the ICP-MS as a "revolution in archaeological provenance studies".

A recent variation of the ICP-MS is the laser ablation ICP-MS or LA-ICP-MS (analytical terminology is almost as replete with acronyms as that for computers!). The laser ablation ICP-MS was first developed in 1985 with several custom-built units operating in 1991. By using a laser such as the Nd-YAG type shown in Fig. 7.8, small (~50 μm) ablation pits can be drilled into sample surfaces with the vaporized ions entrained in an argon gas flow and thence to the quadrapole or magnet-type MS. At the author's home department, both isotopic "lines" for carbonate and sulfur analyses use laser ablation. The ICP-MS can rapidly assess about 70 elements at the ppb level (Kennett et al. 2001).

The key to the ICP instrument, as we have seen, is the RF inductively coupled flame which will ignite a nebulized liquid sample mixed with argon gas that has been injected into it. The torch, diagramatically shown in Fig. 7.7, burns at 8000 °C and ionizes the sample into a plasma. The ionized gas is injected into the mass spectrometer. In the quadrapole instruments, rapid alteration of the voltage across the pole rods allows the transmission of elements of differing mass/charge ratios. In the magnetic sector instruments, the plasma ions enter a curved flight tube where they are separated by their mass/charge ratios. The ion beam can be tightly focused in these instruments allowing the discrimination of masses as close as 0.0001 atomic mass units (amu; Kennett et al. 2001). Laser ablation ICP-MS has been used in the char-

LA-ICP-MS

acterization of Egyptian basalt quarries by Greenough et al. (1999) together with the elemental analysis of clays of Egyptian ceramics (Mallory-Greenough and Greenough 1998) as well as other materials (Gratuze et al. 1993). An attractive feature of the laser ablation instruments is their relatively nondestructive nature, providing one can get the sample beneath the laser. In most cases this is possible. Compared to the nebulized or otherwise digested samples, the artifact remains relatively undamaged (typical damage pits etched by the laser are only tens of microns in diameter (Jarvis et al. 1992).

7.9
Neutron Activation Analysis   NAA

Neutron activation can be used to determine either the absolute or the relative quantities of specific isotopes present in a sample. In practice it is a simple matter to calculate the weight of the element being determined in the sample from the relationship:

Weight of element in sample $\mu g/g =$
  weight of element in standard $\times C_r/C_s$

where C_r is the observed counting rate of the sample and C_s that of the standard measured under comparable conditions.

Nuclear reactions caused by neutrons are of interest in archaeological geology and geochemistry because such reactions are used for analytical purposes to measure the concentrations of trace elements in both archaeological and geological materials. Neutrons produced by fission of ^{235}U atoms are emitted with high velocities and are called "fast" neutrons. For a controlled reaction, the fast neutrons must be slowed down because the fission reaction requires other than "fast" neutrons. Slowing down neutrons requires a "moderator" such as water or graphite with which the fast neutrons can collide without being absorbed. Ordinary water typically serves this purpose in the so-called swimming pool reactors.

Slow neutrons are readily absorbed by the nuclei of most of the stable isotopes of the elements. The neutron number of the product nucleus is increased by 1 compared with the target nucleus, but Z remains unchanged, so the product is an isotope of the same element as the target. The product nucleus is left in an excited state and de-excites by emission of gamma rays. The absorption of a slow neutron by the nucleus of an atom can be represented by the following equation:

$$^{A}_{Z}X(n,\gamma)\,^{A+1}_{Z}Y, \text{ for example, } ^{35}Cl(n,\gamma)^{36}Cl$$

The products of (n, γ) reactions may be either stable or unstable. Unstable nuclides produced by neutron irradiation are radioactive and the product nuclides decay with characteristic half-lives. As a result of slow-neutron irradiation, a sample composed of the stable atoms of a variety of elements produces several radioactive isotopes of these "activated" elements. This induced activity of a radioactive isotope of an element in the irradiated sample depends on many factors, including the concentration of that element in the sample. This is the basis for using neutron activation as an analytical tool.

In routine laboratory practice neutron activation involves: (1) the irradiation of some sample with neutrons to produce radioactive isotopes; (2) the measuring of the radioactive emissions from the irradiated sample; and (3) the identification, through the energy, type, and half-life of the emissions, of the radioactive isotopes produced. Irradiation of a sample with neutrons is carried out with one of three types of neutron sources – reactors, accelerators and radioactive sources. Neutrons produced in nuclear reactors represent the strongest source of neutrons in the sense of producing the most induced radioactivity in the sample. In this case, the sample is directly inserted within the reactor. In the other two types of neutron sources, very often the radioactivity induced within a sample may be too weak to measure when produced by neutron bombardment in ways other than nuclear reactors. The least desirable technique for neutron irradiation is through the use of radioactive sources. Radioactive sources such as Californium do emit neutrons, but others produce alpha particles that are absorbed by materials wrapped around the source to produce neutrons.

The stronger the neutron source, therefore, the more isotopes will reach a detectable level of radioactivity. The measuring of radioactive emissions from the irradiated sources almost always involves the detection of the gamma radiation from the sources. There are two reasons why gamma rays are used. First, of all the radioactive emissions, gamma rays are the least influenced by the structure of a material. A sizable fraction of the gamma rays which escape the radioactive material usually do so with their energy unchanged. Second, since gamma rays are emitted in mono-energetic groups, gamma energies can be used to determine the isotope which has emitted them. The half-life of the nuclide which emitted the gamma rays serves as an additional check on the radioactive isotope created by neutron bombardment. The gamma rays are detected and displayed using devices called gamma-ray spectrometers which exhibit the energies of the detected gamma rays as a series of peaks. The spectrum is calibrated by the use of an energy vs. intensity plot made with reference to gamma rays of known energy. These spectra have a tendency to have relatively sharp peaks which correspond to gamma rays absorption in the detector through the photoelectric effect. These sharp peaks are usually superimposed on and/or interspersed with relatively broad peaks which result from scattering of gamma rays through the Compton effect. Few, if any, investigators use the Compton peaks for interpretive work. Most work is done with the relatively sharp photoelectric peaks which will be spread along the energy spectrum. The height of the photoelectric peak, if it is not superposed on the broad elevated background energy caused by Compton scattering, is directly proportional to the intensity of the radioactivity of the isotope that produces the peak. The photo-peak plot can still be used in quantitative analysis, but is less sensitive because of the presence of background.

Generally, more than one gamma-ray energy spectrum is taken at successive periods of time which are in the order of the half-life of the radioactive isotopes expected. In successive spectra, some series of peaks diminish in height faster than others and in a manner predictable from the half-life of the isotope producing the particular gamma rays. Good examples of this are the isotopes of metallic elements typical of ancient coinage such as silver (Ag) and copper (Cu). Typical isotopes of copper are ^{64}Cu, ^{66}Cu, and those of silver are ^{108}Ag, ^{110m}Ag. For example, silver-108 has a half-life of 2.4 min while that of silver-110m is 253 days. Their respective gamma ray energies are: 636 KeV (^{108}Ag) vs. 412 KeV (^{110m}Ag). NAA, like ICP, requires very little of the artifact, mg-level amounts as a rule, are enough to do the analysis. In this regard NAA can be considered "nondestructive".

7.9.1
Provenance Determination

Provenance studies involve characterizing and locating the natural sources of the raw materials used to make artifacts of clay, stone and metals and thus establish the pattern of trade and exchange (Tite 1992). For ceramics, the use of neutron activation analysis began early on with pioneering work done by Edward Sayre and later with his colleague, Garmon Harbottle, at Brookhaven National Laboratory in the US (Sayre and Dodson 1957; Harbottle 1976). For most elements the ratios of stable isotope abundances are invariant. The relative abundance of elements found in a particular sample is a function of where the sample is found. This makes the identification of prehistoric ceramics and trade goods, such as flint and obsidian, by element composition possible (Bishop 1992; Glasscock 1992; Hoard et al. 1992; Glasscock et al. 1994, 1996; Ambroz et al. 2001; Descantes et al. 2001). Tykot (1996) asserts that NAA together with XRF and increasingly ICP have proven the most successful in determining sources of obsidian.

Ceramic sourcing involves the definition of compositionally homogenous samples of unknown origin (Arnold et al. 1999). The pottery of a particular group of people can be traced through the location of the source of clay used for the pottery and determination of the trace elements in that clay. This is at the heart of what is called "the provenance postulate" which assumes groups of homogenous pottery samples of unknown origin represent geographically restricted sources (Arnold et al. 1999). However, inclusion of temper in the clay before firing might vary the trace elements characteristic of the clay deposit itself. Even when the inclusion of temper creates a problem in the variation of trace elements of the clay deposits, the temper might itself be characteristic (cf. Greenough et al. 1999). The elemental composition of the temper, if sufficiently specific to a particular group, might then serve as a marker of that group regardless of where the clay was obtained.

The digestion and removal of tempering materials such as shell can be done using relatively weak acids without altering the bulk composition of the ceramic clay. If this is done, the neutron activation of the clay can address the question of the clay source without the interference of extraneous, nonsource minerals introduced by the temper. Whatever the specific sample preparation, the application of NAA in ceramic studies has been legend and immensely productive for archaeology. In a recent study of Mayan ceramics, the authors conclude that "Clearly, NAA 'works' in differentiating production communities that use discrete source areas" (Arnold et al. 1999). In numerous other preceding studies, the same can generally be said of the success of NAA as a technique for the study of archaeological provenance.

7.10
Electron Spin Resonance

In a landmark experiment, Stern and Gerlach demonstrated that the electron has the property of spin. As such, it has a component known as the spin angular momentum. The spinning electron behaves like a small bar magnet or magnetic dipole in that it has a magnetic moment defined as:

$$W = -\mu \cdot H$$

Here, μ is the magnetic permittivity (also referred to as permeability) and H is the external magnetic field strength. If an orbital shell is not filled, it will have unpaired electrons in any subshell for which $L \neq 0$. A free atom or ion will have no orbital magnetic moment if it has no resultant spin ($S = 0$).

If a sample of free electrons is placed in a magnet and illuminated with microwave photons such that $h\upsilon$ or $\Delta E = g\beta H$, we have a resonance absorption of energy from the microwave beam as electrons are excited from E_- to E_+ (Fig. 7.10). H is the external field strength as before; β is the Bohr magneton and Landé g spectroscopic splitting factor or "g-value" representing the frequency of the rotation of the electron (Ikeya 1985).

Transitions between these Zeeman levels involve a change in the orientation of the electrons' magnetic moment. In most evaluations, the g-factor contains all the microscopic information that can be obtained from an analysis of the experimental data. Departures from the "free spin" value of g, 2.0023, tell us the species and atomic (crystalline or otherwise) environments within which the unpaired electronics are found. Based on this straightforward information we can rapidly identify elements of archaeological significance such as Mn in marbles (Fig. 7.10a,b).

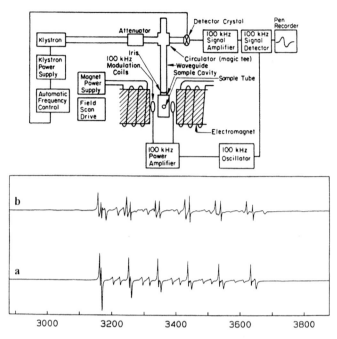

Fig. 7.10. *Above* Typical X-band ESR spectrometer, *below* hyperfine splitting as observed in ESR spectra of two marbles (*a, b*)

Two other features of an ESR spectrum can also give detailed information on the electron wave function and its environment:

1. *Fine structure*: if the total spin of the system involves several interacting electrons, we may have $|S_z| > H/2$ (but always integral or half integral). Then the interaction with microscopic electric fields can split some energy levels even in the absence of a magnetic field. These electric field splittings can also be measured by ESR and serve, for example, to define the local symmetry.

2. *Hyperfine structure*: if a nucleus with a magnetic moment is in the vicinity of an unpaired electron, the effective magnetic field acting on the electron is the sum of nuclear and applied fields $H = H_0 + H_{nucleus}$. The quantity $H_{nucleus}$ depends on the possible quantized orientations of the nucleus, S, that each ESR line splits into several components. This can be a valuable diagnostic tool (e.g., in studies of ancient marbles) where the hyperfine spectrum is a direct "fingerprint" of this element, thus aiding in differentiating marble facies (Fig. 7.10).

ESR works only on paramagnetic materials, i.e., those with unpaired electrons. As we shall see in the following section on magnetic susceptibility, most solids have their electrons in closed (paired) shells and are dia-or nonmagnetic as a result. Because paramagnetism is in complete alignment with the magnetic field, the net magnetic moment is significant in most cases.

To induce paramagnetic behavior in a material it is placed in an ESR spectrometer which is an apparatus (Fig. 7.10a) with an electromagnet having a sample holder between the poles at the end of a brass microwave (radiation) wave guide. The sample is placed between the magnet's poles and microwave radiation at a constant frequency, ca. 9.3 GHz, and a spectrum is "swept" between the levels of magnetic field strength, say 100 Hz. Recalling the basic equation for ESR spectroscopy, the sample will absorb the microwave radiation (resonance) at some value of the g-factor.

To date, in archaeology, ESR has been increasingly used in the last two decades for dating purposes (Wagner 1998), whereas in mainstream chemical analysis it has been used for studies of the properties of numerous compounds and materials. In some recent cases, ESR has been used to evaluate heat treatment of flints (Dunnell et al. 1994) and to provenance marbles (Baïetto et al. 1999).

7.11
Magnetic Susceptibility

The impression of a magnetic field on a material will induce a magnetic response. The nature of that response is a direct result of the electronic structure of the material's constituent atoms. We have seen that the source of a paramagnetic response, ESR, is the unpaired orbital electrons in the valence (outer) shell of the atom. Diamagnetic magnetism results from the precession or distortion of electronic orbitals in a magnetic field. All materials are diamagnetic (Verhoogen 1969). Certain crystalline materials have overlaps of the electronic orbitals of their constituent atoms. This "coupling" of the orbital magnetic moments results in coupling in either parallel or antiparallel directions (Fig. 7.11) producing a strong spontaneous ferromagnetic response such as magnetite. The antiparallel response is found in an antiferromagnetic material. Strong magnetism is observed in ferrimagnetic materials. The response is qualitatively similar to ferromagnetics. The difference between the two is that the magnetic ions in the ferrimagnetic material align more readily in an antiparallel direction. The ferrimagnet's crystalline structure is more complex than the ferromagnet with sublattices that create the parallel or antiparallel magnetic response. The overall response, in an applied magnetic field, is either ferri- or antiferrimagnetic depending upon the material. Spinal and garnet are two ferrimagnetic minerals.

To measure magnetic susceptibility we place the material in the magnetic device – a balance, bridge or meter. The susceptibility measurement is obtained by measuring the net magnetic moment, I, per unit volume or mass in the applied magnetic field, H. The equation to do this is as follows:

$$I = \kappa H$$

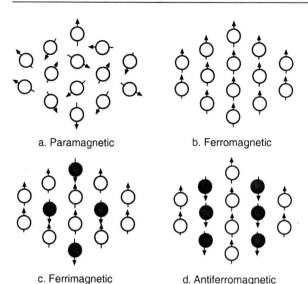

Fig. 7.11. The magnetic moment vectors in various forms of magnetic materials

a. Paramagnetic

b. Ferromagnetic

c. Ferrimagnetic

d. Antiferromagnetic

where κ is the susceptibility. For paramagnetic materials κ is positive. For diamagnetic materials it is negative; for ferro-, ferrimagnets it will be strongly one or the other. The magnetic susceptibility measure is dimensionless which means it has no units, it is simply a number. Except in devices like the Guoy magnetic balance, the paramagnetic response will be ferro-or ferrimagnetic, in either mode, ±.

The induced magnetism of common ferro- or ferrimagnetic materials is shown in Table 7.1. As a general rule rocks, particularly igneous ones are more magnetic than sediments or soils. Rocks can range as high as 10^{-1}–$1/g$ while sediments are more commonly in the range of 10^{-5}–$10^{-6}/g$ (Table 7.2).

Magnetic susceptibility measurement devices are the simplest of all magnetic instrumentation. The measurement of magnetic susceptibility for the study of chemical species in solution has a long history. The body of magnetic theory that supports these inquiries are some of the better understood principles in recent science. Only recently have earth scientists – geophysicists mostly – and some archaeological geologists taken an interest in the use of susceptibility as an analytical tool. In the latter case it is because the magnetic susceptibilities of soils reflect their parentage (Sharma 1997). Soils derived from igneous and metamorphic rocks, notably those of more mafic character, have greater amounts of magnetite and appreciable magnetic susceptibilities (cf. Table 3.2). Soils of a highly organic nature, as can be found in archaeological settings, can have elevated amounts of maghemite, which can be seen to be significantly stronger than hematite. Because of the production of both maghemite and the increase in remnant magnetization due to fire such as in hearths and conflagrations (Le Borgne 1960; Mullins 1974),

Table 7.1. Magnetic properties of soil minerals

Mineral	Formula	Magnetism
Magnetite	Fe_3O_4	Ferrimagnetic
Hematite	$\alpha\text{-}Fe_2O_3$	Antiferromagnetic
Maghemite	$\gamma\text{-}Fe_2O_3$	Ferrimagnetic
Goethite	$\alpha\text{-}FeOOH$	Antiferromagnetic
Lepidocrocite	$\gamma\text{-}FeOOH$	Antiferromagnetic
Ilmenite	$FeTiO_3$	Antiferromagnetic
Pyrrhotite	Fe_7S_8	Ferrimagnetic
Pyrolusite	MnO_2	Antiferromagnetic

Table 7.2. Typical measured mass magnetic susceptibilities of common primary and secondary minerals

Mineral	$\times 10^{-6}$ (cgs units)[a]
Corundum	−0.34
Orthoclase, calcite	−0.38
Quartz	−0.46
Dolomite	+0.9
Muscovite	+1.0–12
Biotite	+12–52
Amphiboles	+13–55
Pyroxenes	+3.0–75
Hematite	+20
Siderite	+93
Magnetite	up to +800,000
Lepidocrocite, goethite	+42
Maghemite	up to +30,000
Kaolinite	−1.5
Montmorillonite	+2.2
Nontronite	+69
Vermiculite	+12

[a]To convert from cgs units ($cm^3 g^{-1}$) to rationalized SI units ($m^3 kg^{-1}$), multiply values by 4×10^{-3}.

the anthrosol has magnetic variations just beginning to be appreciated in archaeological geology (Marmet et al. 1999).

Tite and Linington (1975) suggested that magnetic susceptibility of anthropogenic sediments and soils could be viewed as a proxy for past climates. Magnetic susceptibility is viewed as a result of paleoclimatic controls such that when pedogenesis is high, as during interglacials, so is magnetic susceptibility (Kukla et al. 1988; Ellwood et al. 1994, 1995). Studies of deep sea sediments have shown a close correlation of magnetic susceptibility to mineralization and grain size (Bond et al. 1993; Kennett et al. 1994; Seidov and Maslin 1999; Maslin 2000).

7.12
Cathodoluminescence Microscopy

Cathodoluminescence (CL) is the emission of photons in the visible range of the electromagnetic spectrum after excitation by high-energy electrons in an electron microscope. It is the most important optical phenomenon used by all electron microscopists (Heard 1996). It is an effect coincident with several other forms of stimulated emission such as primary and secondary X-ray emission, and secondary and backscattered electrons. CL results from electrons in the material being excited into the conduction band and the subsequent recombination in luminescent centers at the valence band, resulting in the release of visible spectrum photons readily seen by photomultipliers (Ramseyer et al. 1989; Barbin et al. 1992; Fig. 7.12). The energy difference between the conduction and valence levels, only a few eV, determines the wavelength of the luminescence.

Beginning in 1965, CL became of interest to geologists. The hot-cathode CL microscope consists of a hot tungsten cathode under a vacuum, the acceler-

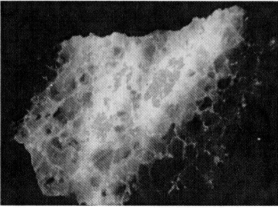

Fig. 7.12.
Cathodoluminescence of marble. *Above* Marble thin section, *below* thin section in luminescence (Images country Prof. L. Barbin)

ating voltage ranges between 2.5 and 50 KeV, and current densities are between 5 and 10 nA/mm^2. A monochromator with a spectral range of 350–850 nm is fitted between the cathode and viewing optics. CL microscopy is practised in two differing procedures: hot-cathode CL and low-temperature CL. To date, the hot-cathode method has been most used in archaeological geology.

In archaeological geological studies, CL has made its biggest impact in the provenance studies of white marbles. CL studies have divided white marbles into three major families based on luminescent color . Calcitic white marbles have a dominant orange or blue luminescence, while dolomitic marbles show a red luminescence. By combining CL with grain size, texture and, most important, stable isotope analysis, finer distinctions can be obtained between marbles (Barbin et al. 1991, 1992). The excited mineral, calcite, dolomite, etc., produces different colors of luminescence that originate from various impurities in the crystal lattice. The lattice defect in carbonates most used for CL is that of Mn^{2+}, zircons with dysprosium (Dy^{3-}) and europium (Eu^{2+}), apatites with Mn^{2+}, Dy^{3+}, and samarium (Sm^{3+}), and fluorite Eu^{2+} and Eu^{3+} centers.

The orange luminescence family contains most of the calcitic marbles from quarries such as Mount Pentelikon and Thasos in Greece: Naxos, Paros, and Pteleos in the Cyclades; Carrara and Lasa in Italy, and Dokimeion in Turkey. Blue luminescing marbles are less easy to characterize because the cause of the CL coloration is not established. The blue emission is observed in marbles with manganese present in the calcite at levels below 5 μg/g. Red luminescing marbles are exclusively dolomitic (>50% dolomite). Representative marbles include those from Carrara, Italy; Marmara, Turkey; Paros and Naxos, in the Cyclades (see Table 7.3).

CL can be used to characterize surface texture and composition of minerals in geological and ceramic materials. Traditional optical microscopes can be fitted with CL stages such that low-voltage electrons are produced which can produce CL in materials under examination. While useful it is best to use the broader range and higher accelerating voltages produced by SEMs for high resolution and precise CL measurements. These devices can readily measure both color and intensity rapidly and quantitatively.

Recent research (Yacobi 1994; Heard 1996) suggests the combination of CL and other forms of spectroscopy (BSE, EDS, etc.) can provide important information on the composition and uniformity of ceramic materials such as clays, cements and silicate glasses that readily luminesce. For instance, the temperature of heating or firing can be examined by the presence or absence of CL related to monoclinic phases of oxides such as CaO. Heating to a degree that eliminates the monoclinic phase removes all CL. In addition, quartz grains of relatively uniform density seen in BSE images have intensities related to grain growth and boundary behavior. Such behavior should be related to phase changes in quartz found in ceramics as either temper or as amendments to the mineral suite used in the particular ceramic such as porcelains.

Table 7.3. Cathodoluminescence and other characteristics of selected white marbles. (Adapted from Barbin et al. 1991)

Locality	Texture	Grain size max. (mm)	Mineralogy	Cathodoluminescence		
				Color	Intensity	Int. Distr.
Thassos-Aliki	Homeoblastic	1.4–5.0	Calcite	Orange	Medium – strong	Homogeneous
Thassos-Vathy	Slightly cataclastic	0.4–1.8	Dolomite	Red	Medium – strong	Homogeneous
Paros-Stephani	Homeoblastic – weakly porphyroblastic	0.3–1.5	Calcite	Dark blue to blue-pink	Faint	Homogeneous
Paros-Chorodaki	Slightly Porphyroblastic	0.5–3.7	Calcite	Reddish-brown	Faint – medium	Homogeneous
Pentelikon	Slightly porphyroblastic	0.3–1.0	Calcite	Orange	Faint – strong	Patchy
Usak-Kavacik	Highly porphyroblastic	0.4–2.0	Calcite	Orange-brown yellow	Faint – strong	Zoned
Carrara	Homeoblastic	0.06–1.3	Calcite	Orange	Medium – strong	Homogeneous
Afyon	Porphyroblastic	0.5–1.4	Calcite	Orange to blue	Very weak	Patchy
Marmara	Porphyroblastic	0.2–3.6	Calcite	Blue	Very weak	Homogeneous
Naxos-Flerio	Porphyroblastic	0.7–7.5	Calcite	Orange	Medium	Homogeneous

Picouet et al. (1999) have used CL in the study of quartz, feldspar and calcite grains in pottery as a potential tool for provenance studies – the separation of late Neolithic–early Bronze Age pottery, from archaeological sites in western Switzerland and eastern France. Akridge and Benoit (2001) have examined the CL in chert artifacts from the Ozark Mountains of the central USA. These cherts showed predominately orange CL with an occasional blue CL observed. As chert is principally considered a microquartzite (Gerrard 1991), the CL properties are those of quartz. In a study of geological quartz, Walderhaug and Rykkje (2000) suggest that the crystallographic orientation of the quartz grains may have a significant effect on observed CL.

7.13
Infrared Spectroscopy

Single atoms do not emit or absorb infrared radiation (Goffer 1980). Molecules absorb infrared frequency radiation in the 2.5–16-μm-wavelength range. As a consequence, nearly all molecular compounds show some degree of absorption in the IR part of the electromagnetic spectrum. It is best used as a qualitative rather than a quantitative technique.

As such it is the most widely used spectroscopic method for analysis of organic compounds and mixtures. Double-beam IR, NIR (near infrared) and FTIR (Fourier transform IR) have come to dominate this area of spectroscopy and as a result of the new chemometric computing methods used with these instruments, the quantitative capabilities have increased dramatically. Field-hardened FTIR units have been deployed at Kebara Cave in Israel (Weiner et al. 1993).

The use of FTIR at the Kebara site illustrates the potential of the use of IR spectroscopy in archaeology. The researchers were able to detect the presence/absence of calcium hydroxyapatite or dahllite – in cave sediments. FTIR was thus able to map the concentrations of bone and to discriminate these areas from those of bone dissolution, due to diagenesis and areas of partial bone preservation (Weiner et al. 1993). As mentioned above in the section on mass spectrometry, FTIR has been coupled to gas chromatography with potentially interesting results for archaeology.

7.14
Instrumental Geochemical Techniques
and Their Availability to Archaeology

Modern instrumental geochemical analyses are available on a fee basis throughout the world. In most research universities, many, of the methods discussed in this chapter can be found. The important thing for the archaeological investigator to determine is which technique is best able to assist in evaluating the specific research question. By way of illustration, Shackley

(1998) has discussed just this question with reference to archaeological obsidians. In the late 1960s and early 1970s, XRF and NAA were increasingly used to deal with issues of obsidian exchange and interaction. By the 1980s archaeologists in every part of the world, where obsidian was used were engaged in geochemical studies of this material. The range of instrumentation has expanded as well to include ICP and PIXE, together with the newer models of SEM. Shackley's comparison of the three top techniques in obsidian analyses illustrates points the archaeologist need to consider in choosing an analytical technique.

NAA endured into the late twentieth century as the geochemical method perhaps best known to archaeologists. Part of this was due to the almost missionary zeal of some archeometrists in proselytizing their archaeological fellows for its use in provenance and characterization studies from ceramics to obsidians. As Shackley and others (Glasscock 1991; Neff and Glasscock 1995) point out, NAA does have two primary shortcomings – it is a destructive technique and it cannot analyze samples for barium (Ba) and strontium (Sr). While NAA labs pride themselves in the small portions (~mg) ordinarily used in their analyses, the point is that the artifact, if indeed it is such, must be broken however small the damage. In addition, the material can be radioactive, depending on the mineralogical nature of the sample, for years. Finally, realistic or not, the general public has developed a general fear of radioactivity. While more scientifically versed than most, archaeologists can and do share these concerns. For museum specimens and artifacts that are subject to repatriation (see the discussion in the last section of this chapter), NAA is not the method of choice.

XRF and NAA are remarkably comparable in their results for obsidians, particularly in the mid-Z and part of the high-Z region for more than 17 elements analyzed in common. This comparability of analytical results now extends to ICP-MS and PIXE as well. What this means to the archaeologist is that any laboratory employing XRF-EDS, XRF-WDS, NAA, PIXE or ICP-MS will provide valid and comparable results. To a great degree, Shackley's conclusions concerning the applicability of analytical techniques to obsidian studies can be extended to the great majority of archaeological materials – either inorganic or organic in nature.

When the researcher is not comfortable in his/her own knowledge, consultation with either specialists or commercial laboratories can help. This is not to say competency in assessing the relative merits of instrumental geochemical techniques will not increase. Today, it is common for the field archaeologist to select, collect and submit radiocarbon samples without consultation unless it is to decide whether to choose the conventional, beta-counting radiocarbon technique or the somewhat more esoteric atom-counting AMS method (Fig. 7.13).

It is also not uncommon for more archaeologists to subscribe to or be familiar with international specialty journals on archaeometry and archaeological science such as *Archaeometry*; *Revue d'archéologie*, and the *Journal of*

Fig. 7.13. The 500 kV NEC Model 1.5SDH-1 Pelletron AMS system, Center for Applied Isotope Studies, University of Georgia. (Photograph by author)

Archaeological Science. Add to this trend the gratifying addition of more technically oriented articles in more traditional archaeological journals such as *Antiquity, Journal of Field Archaeology,* and *American Antiquity.* Articles on the geochemical analysis of English ceramics have recently expanded the format of the journal *Historical Archaeology.* All of this reportage of instrumental geochemical analysis in publications commonly read by the archaeological community increases both knowledge and awareness of their usage on commonly held research questions.

Archaeological science has presented a veritable cafeteria spread of both geochronological techniques – radiocarbon, uranium series, electron spin resonance (ESR), optical-and-thermoluminescence (OSL, TL) – and geochemical methods. As archaeological geologists it is appropriate to increase the application of both these dating and analytical tools in archaeology. In a personal survey of brochures and catalogues from commercial analytical laboratories preparatory to this volume, I noted offerings of routine assays of 66–70 elements by ICP-OES/MS and 29 elements by NAA. In the XRF of common metallic oxides, 11 were routinely analyzed by the WDS method. As we have seen in the discussion of obsidian, the University of California's Hearst museum routinely analyzes 17 elements. Figure 7.14 summarizes the analytical methods and their detection limits. The latter analytical parameters occur in relatively higher concentrations ($\mu g/g$–%) while trace elements and rare earth elements (REE) are more often found in $\mu g/g$–ng/g concentrations. In the decision on which geochemical protocol to use, the X-ray and electron probe techniques provide relatively inexpensive and quick analyses of the higher concentration elemental parameters. For parameters that occur in lower frequency and concentration the ICP and particle-based methods (NAA, PIXE) provide equal analytical access for archaeological inquiry. Almost any archaeological material is analyzable – ceramic, lithic, faunal, botanical, sediment, soil. If the technique requires

11.29

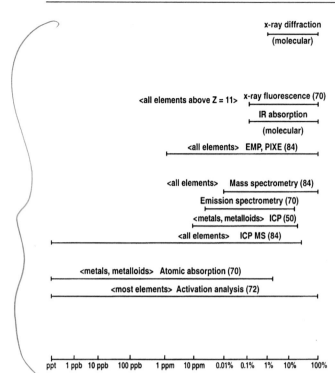

Fig. 7.14. Detection limits for major analytical techniques

digestion of the sample and is thus destructive, the compensation is that relatively small amounts are needed. (mg–g levels). In one study of archaeological sediments which we recovered in amounts of only a few grams, it was gratifying to find that only milligram amounts were needed to assess elements and minerals present. At the end of the study involving textural, geochemical and modest geochronological analyses, we were left with some remainder of all our samples. Given the analytical power and precision of conventional instrumental geochemical techniques, there are few items of archaeological interest that cannot be examined with little or no sacrifice of that material. Even where the destruction of any part of an artifact is unthinkable, the non-destructive whole sample methods-XRF, NAA, EMP/SEM, and LA-ICP-MS are available. Figure 7.14 summarizes the various techniques discussed in this chapter.

Having reported so glowingly to the reader it is instructive to recall my own frustration with the loss of a singular piece of early horticulture in the US – a small, carbonized corn cob – in a desire to obtain chronological information with the AMS-radiocarbon technique. One of the major "selling" points of AMS is the proportionally small sample needed to calculate the age of the artifact/item. While it is true that μg samples can be used, most dating laboratories will insist on the sacrifice/digestion of as much of the sample as possible. Theory and practice are often at odds in the specific analytical case such as that of the small corncob. In that case the results were: (1) a reliable age determination with direct evidence for early corn horticulture in the American southeast and (2) the loss of a singular, exhibitable example of that early plant domestication.

7.15
The Struggle over Bones

As we enter the third millennium the practice of archaeology routinely includes the instrumental analysis of the materials – artifactual and contextual. Generally, these studies occur after the fieldwork is complete, but more of these analyses may be done during the course of field research such as the use of FTIR on skeletal material from Galilean cave sites by Weiner et al. (1993), Stiner and Goldberg (2000). Field-hardened spectroscopic instruments coupled with laptop portable computers can accomplish high precision analyses. As more demands are made for rapid turn-around, compliance type of archaeological research such as CRM (cultural resource management), the salvage/rescue-type archaeology of the late twentieth century United States and elsewhere, the archaeological researcher will require such instrumental support. This is becoming more the case in modern study of ancient burials in the US and Israel (Morell 1995). Williams-Thorpe et al. (1999) have reported the use of a field-portable XRF device which could have great utility in in situ studies.

In situ measurements, beyond morphometric parameters, of aboriginal skeletal finds may become the only acceptable method of obtaining chemical data on human bones and teeth. As a scientist of both Euro- and native-American descent, the author personally supports expeditious, nondestructive tests of human skeletal material, but recognizes the absolute necessity for respect and the responsible regard for the rights and beliefs of the modern descendants of prehistoric aboriginal peoples. Nothing has impacted the practice of Americanist archaeology in the late twentieth century United States more than the passage of the Native American Graves Protection and Repatriation Act of 1990 (NAGPRA). The impact of this legislation is still being measured. Suffice to say many other minority peoples are watching and the US has not been the only locus for the conflict over human remains (Morell 1995).

The most prominent battle between native rights and scientific interests has been that of the so-called Kennewick Man (Preston 1997; Feder 1997). Found in 1994, by accident, along the banks of the Colombia River in the US state of Washington, the nearly intact skeleton is that of a 9000-year-old male with a stone projectile point embedded in the pelvic bone and cranial features unlike those of most aboriginal archaeological examples. After an extensive 5-year tug-of-war in the US courts, a compromise was reached allowing instrumental tests – DNA, isotopes, etc. – to be made on the skeletal remains.

At the time of writing these tests have not been published, but preliminary statements indicate this ancient American, first considered Caucasoid on the basis of cranial morphology, may share a trans-Pacific genetic heritage, adding a new dimension to the question of New World settlement. Whatever the origin of the Kennewick Man, the precedents set in the court-mandated scientific studies bespeak a more responsible and expeditious approach to the study of future finds. It is no longer permissible to simply excavate human remains, leisurely analyze and then curate them in perpetuity as has so often been the fate of thousands of archaeological finds in the US and elsewhere. Science can be served, but cultural demands must be recognized and balanced in the process (Renfrew and Bahn 1996; Fagan 1999).

7.16
A Closing Quote

Z.A. Stos-Gale (1995), in her review of isotope archaeology, quotes Gunter Faure's introduction to his 1977 volume on isotope geology and it seems as appropriate in this context as hers: "The time has come to introduce this subject into the curriculum in order to prepare geologists in all branches of our science to use this source of information. Although the measurements on which isotope geology is based will probably continue to be made by a small, but expanding number of experts, the interpretation of the data should be shared with geologists who are familiar with the complexity of the geological problems." Stos-Gale suggests perhaps the time has come to substitute the word "archaeology" for that of "geology".

8 Statistics in Archaeological Geology

8.1
Introduction

Waltham (1994) considers statistics the most intensively used branch of mathematics in the earth sciences. His textbook, along with that of Robert Drennan, *Statistics for Archaeologists* (1996), are excellent introductions to statistics appropriate for archaeological geology. Following Waltham, the definition of a statistic is simply an estimate of a parameter – mass, velocity, dimension, etc. – based upon a sample from a population. Unless that population is composed of a relatively small number of items or objects then it almost certain that estimates must be made of the populations using independently drawn samples. As archaeology is largely a study of a population of "things", reliable estimates of these collections are best made using statistical techniques. These techniques include parametric measures of central tendency and dispersion such as the mean, the standard deviation and the variance.

Samples are observations of larger populations. For geological samples they are the actual rocks and measurements. Taking on specific variables – the parameters such that a set number of observations of these variables make up a sample set. Dispersion is quantified and compared by statistical tests such as simple regression analysis on two variable or bivariate data. In addition, a central question for archaeological questions of interest is that of "how big" a sample set is needed to adequately test a difference in sample set means.

In earlier chapters we have already alluded to statistical issues in our discussion of scale and sample size (Chap. 4); particle size analysis (Chap. 5); sample size and regression or line-of-best fit (Chap. 5). These are very common issues in both archaeology and geology. The justification of including an introductory chapter on statistics stems from the real need to evaluate one's data for insights that are not that readily apparent. It is a necessary part of any rigorous science. The popularity of statistics in archaeological geology can be appreciated by a simple survey of any major journal wherein these techniques commonly appear. For example, in just one recent issue of the *Journal of Archaeological Science* 8 of the 12 articles used statistical techniques of one form or another to present their data. While less common in journals of archaeology per se, even there one observes the increasingly frequent use of statistics. A 1999 issue of *American Antiquity* alone produced an impressive range of statistical methods ranging from regression analyses, Bayesian estimations, diversity measures, to spatial and multivariate analyses in *all* five of the major research reports of that issue. Neither of these examples is unusual.

The focus of this chapter will be to list and briefly describe commonly used concepts and techniques in *data analysis*. It is beyond the scope of this book to attempt anything more than this. As well as basic concepts, we shall look at ways used in the examination of large data sets that involve more that two variables – the so-called multivariate analytical methods. A few brief examples from archaeology and modern DNA studies of early hominids will be used to illustrate the techniques or topics under discussion. The reader should take away from this chapter a sense that statistics offer ways to solve problems that may not done otherwise.

8.2
Descriptive Statistics

Descriptive statistics consists of the methods for organizing and summarizing information in a clear and organized way. In contrast, inferential statistics consists of the methods used in drawing conclusions based on information obtained from a sample of a population (Weiss and Hassett 1993). Modern statistics is principally that of inferential statistics. We begin our discussion with descriptive statistical measures.

8.2.1
Scales

One area that will not be examined in any depth is that of nonparametric statistics or the techniques used in the evaluation of issues using ordinal or nominal scaled data. One important technique is that of *chi-square* (χ^2) and that will be discussed in some detail. In addition, many data sets have nonnormal distributions. These either have to be transformed to normal form or dealt with using nonparametric tests such as chi-square. With these caveats stated, we shall nonetheless begin the discussion with these and other types of scales.

Scale types include: (1) nominal, (2) ordinal, (3) interval and (4) ratio. The types of scales differ on how the observations that make up a sample relate to one another. The simplest relationship between observations is that of the nominal scale. The nominal scale accounts only for identity such as "type A", "type B", etc. That is to say it is "categorical", nonranked data. An example would be rock or pottery types. Within "type A" there may be ten observations; "type B" may have five observations. Presence–absence data are nominal scaled.

For ordinal scales ranking is implicit although the ranks may not be of equal magnitude, such as the Moh's hardness scale (cf. Chap. 6). Observations within ordinal classes have an order. Grade categories are ordinally scaled, e.g., A,B,C,D or F imply rank within the grade categories of an A→F order. Ranking is relative just as is magnitude.

Most observations in what is termed parametric statistics are interval scaled. Interval scale data improve on ordinal data as this scale has the continuum of integers divided as equal whole or decimal numbers. Size is scaled along interval scales as is temperature and dates. Interval data are arithmetic such as $2 + 2 = 4$; $2 - 1 = 1$.

Ratio scale data are typified by geometric or progression relationships. For instance $2 + 2 = 4$; $8 \div 2 = 4$ are ratio scale data. The distinction to be noted is that when one converts between interval scales addition and multiplication is necessary. When one converts between ratio scales (e.g., length in centimeters to inches), only multiplication is necessary, hence the term ratio. The ratio scale has an absolute difference in observations; it has a natural, non-arbitrary zero point. Measurements of mass, length, time, velocity, etc. are scaled on a ratio basis.

8.2.2
Variables

Mention should be made of the important distinction between the characteristics of the observations we collect as data. Characteristics of the members of a population vary and are termed *variables*. The two kinds of variables are either: (1) continuous and (2) discrete. An analogy from human genetics illustrates this difference – the gene(s) for human skin color are basically continuous – the wide range in human skin coloration from white through brown to black – while the gene for human eye color, and that for blood groups, is relatively discrete, e.g., blue, brown, or gray/green; A, B, AB and O.

The continuous variable may assume any value within a set range such as fractional numbers such as 30.5, 63.4, 73.5 or 75.1. By comparison, the discrete variables would take on only whole number values such as 30 coins, 25 rocks, etc. One normally does not refer to fractional values of coins such as 30.5 "dimes", but it is perfectly appropriate to describe height as 30.5 cm or mass as 125.25 g. For populations the values of variables are parameters; for samples they are statistics. Measurements on variables are observations or data. Sets of data drawn from a population of variables are samples – the business of statistical inquiry.

8.2.3
Frequency Distributions, the Normal Distribution and Dispersion

Table 8.1 lists the relation between types of scales and their various statistical measures of central tendency.

Table 8.1. Scales and statistical measures

Scale types	Measures of central tendency	Measures of dispersion	Measures of association	Tests of significance
Nominal	Mode		Chi square	
Ordinal	Median	Percentiles	Rank correlation	
Interval	Mean	Average and standard deviation	Correlation ratio	t-test, f-test
Ratio	Geometric mean	Average and standard deviation	Correlation ratio	t-test, f-test

In this table we observe that mode and median are calculated for nominal and ordinal data, respectively, but one may use them for interval and ratio scales as well. In many cases they are appropriate measures of central tendency. Under the tests of significance, chi-square examines the dependence between characteristics or "types" in a population. How well they might be associated is contained in the value of the chi-square statistic – if the calculated chi-square value exceeds the tabulated value, at the predetermined level of significance, then the parameters are dependent or associated. Likewise, the interval and ratio scales examine central tendency – dispersions – and association is examined by correlation coefficients and comparisons between sample means are made using the t- and f-tests. These will be discussed further in the following sections.

Even in this world of spreadsheet software and the graphic presentation of data, the need for an extended description of frequency plots exists. The frequency plot-bar graph, pie chart, etc. shows, in graphical form, how often a particular variable is observed. Figure 8.1 is such a plot. Note the variables are discrete in nature. A hidden danger of the ease with which software makes the production of these presentations is that of appropriateness. Frequency data should be displayed with bar charts, but X–Y data should be shown using X–Y plots. Data visualization is the first step in data analysis, therefore, care should be exercised when applying software to data. This injunction extends from statistical packages included in spreadsheet programs to the more sophisticated forms like SPSS (Statistical Package for the Social Sciences).

A plot of a set of continuous variables will produce a similar looking graph, but since the data are continuous, intervals must be chosen within the continuum. An example is that of a frequency spectrum such as encountered in a conductivity meter's data set. The frequency spectrum can be displayed continuously from, say, 300–19,000 Hz. Values within this range can be measured and displayed as continuous data. In actuality, most conductivity data are displayed within set frequency intervals such that only values at specific frequency settings are plotted such as 300, 3500, 8250, 19,000 Hz, etc. Counts or values of either discrete or continuous data produce histograms.

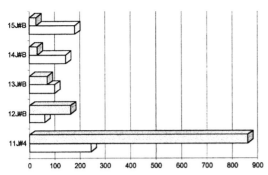

Gray's Reef sediment analyses - Bar chart showing potassium(dark) and calcium (light) values determined by x-ray fluorescence (XRF).

Fig. 8.1. Typical bar chart (histogram) plot of single parameter data

8.2.4
The Normal Distribution

A histogram or the frequency observations of variables is often confused with the probability distribution. The latter are histograms of the probability of a variable and not its observed frequency. One of the most common probability distributions is the normal distribution. The normal distribution can represent either discrete or continuous variables. The great German mathematician, Carl Friedrich Gauss, in the context of determining the frequency of observable stars in a quadrant of the night sky, developed the basis for the normal or "Gaussian" probability distribution. The importance of the normal distribution lies in its central role in statistics. Because it introduces probability, it lies within the province of inferential statistics. A major assumption of normal distribution is that it characterizes the frequency of occurrence of most continuously distributed populations such as Gauss' stars. The normal distribution is graphically familiar as the normal or bell-shaped curve (Fig. 8.2). Key mathematical properties of the normal curve are:

1. The area under the curve is, by definition, 1
2. The measure of central tendency is the mean, which is by definition, zero
3. The curve is perfectly symmetrical about the mean.
4. The standard deviation, along with the mean, completely defines the normal curve
5. The mean, mode and the median *all* have the *same* value.

The normal distribution is a statistical construct which may or may not apply to any given data set. There are other types of probability distributions of discrete variables that are important to statistics such as the Binomial distribution, the Poisson distribution, and the Negative Binomial distribution. The

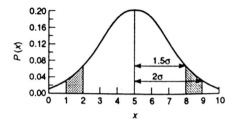

Fig. 8.2. The normal distribution. The probability $P(x)$ of occurrence of any observed values from 1 to 2 and 8 to 9 ("two-tailed") is shown in *shaded areas*. The range of values in this example is 0–10 with the mean ($\bar{x}$) = 5. 2, *1.5σ* indicate the "standard deviations" from the mean. (Adapted from Waltham, 1994, Fig. 7.7, p. 119)

most important of these is the binomial distribution. It is the most widely used discrete probability distribution. By its name it implies that it is used in the study of two variables. Like the normal distribution and the other types, it has an area of 1. However, a mean of zero is singular to the normal distribution alone. Other distributions where the mean, mode and median are not the same and display *skew*. This parameter is commonly calculated in sediment textural analyses typical of those outlined in Chapters 4 and 5. The skewness of the grain size or other particulate distribution, e.g., phytoliths, can be instructive. Another way of thinking of skew is in terms of the symmetry of a distribution. The normal distribution is symmetrical with no skew. Other distributions – Poisson, Gumbel, Pearson Type III, etc. – can, and typically will display, nonsymmetrical curves. In fact, the Poisson is recognized by its sharp skew to the left axis and hence tells one that the sample population can be described thus.

It is the Normal distribution that is most commonly assumed to underlie most phenomena. This is formalized in the Central Limit Theorem. The central limit theorem states the means of samples from a population have a normal distribution provided the sample size, n, is large enough. Note that the central limit theorem does not require the sampled population to be normally distributed. This is conceptually important as it allows inferential statistical tests, such as the t test discussed below to be used on nonnormal data.

8.2.5 Central Tendency

Table 8.1 lists the measures of central tendency appropriate for variously scaled data. The normal distribution's most common measure of central tendency is the mean. It is used to measure data on interval or ratio scales. Here, the mean may not be such a good choice particularly if the distribution is asymmetrical. In this case, perhaps the mode – that data value which occurs most frequently – is a more representative measure. The measure of central tendency for ordinal data is the median. In the normal distribution whose mean is zero, so the median is likewise (see above; Hampton 1994). Unlike the mean, the median is some value that has an equal number of items above and below its value. This is to say that the data value that is exactly in the middle of an ordered list such as in the case of the following seven values – 2 2 2 3 4 4 5; "3" is the median value. For nominal scales, the measure of central tendency is the mode or the most frequently occurring item in the data set. In a set of 10 flint artifacts, 8 basalt artifacts and 5 quartzite artifacts the mode is "flint artifacts".

There is no simple rule for choosing one of the three – mode, median or mean – to measure central tendency. For instance, if the data are bi- or tri-modal then using either the median or mean seems to be a poor choice, whereas the mode is more representative of central tendency in these cases.

Another example could be that of cathodoluminescence observed in a suite of marbles such as we have seen in Chapter 7 (Table 7.3). We see that the researchers have reported the "mode" for the CL color for the marbles, e.g., Thassos-Vathy, a dolomitic marble, is "red"; Carrara marble is "orange", etc. Had the researchers quantified the CL in interval terms, then we might use either the median or mean to describe the "intensity" of the "red" or "orange". Be this as it may, it is the arithmetic mean that one encounters most often in parametric statistics and this what we shall discuss now.

8.2.6
Arithmetic Mean and Its Standard Deviation

The mean of the overall population is symbolized by μ, whereas the sample mean is denoted as $\bar{x}$. It is calculated as follows:

$$\bar{x} = \Sigma x / n$$

or the sum (Σ) of the x observations of a variable divided by n observations. The standard deviation, σ is the mean squared difference between an observation and the mean. The calculation of the standard deviation requires that one must first determine the total variation or variance (s^2) within the sample. The population variance, σ^2, is often illustrated in textbooks, but rarely used. To calculate s^2, the sum of squares (ss) is first calculated. The formula is:

$$ss = \Sigma (x - \bar{x})^2$$

Here, the mean is subtracted from each value of the data set. These differences are then squared and summed. The parameter obtained is divided by n, the number of observations.

$$\sigma^2 = ss / n$$

This is the population variance which can rarely be determined so one uses $n - 1$ in the denominator to determine s^2 or simply: $s^2 = ss/n - 1$. The standard deviation is simply the square root of the variance.

In the normal distribution, the standard deviation is mathematically the point of maximum inflection tangent to that normal curve. Under the normal curve one standard deviation represents 34.13% of the total area under the curve. This is for one "tail" or side of the curve, so for the total curve, one standard deviation represents 68.26% (Fig. 8.2). Typically, the reporting of a radiocarbon date is at one standard deviation from the mean age.

8.3
Inference – Hypothesis Testing, Types of Error and Sample Size

8.3.1
Hypotheses and Testing

In the scientific method, the approach to evaluation of phenomena is through the use of hypothesis testing. Statistics – those which we term *inferential* – aid in the evaluation of hypotheses. A hypothesis is simply a statement that something is true. As such this statement can be tested. The *null* hypothesis is that which is actually tested. An example of this method can be illustrated by use of the CL of marbles. We can assert "all Carrara marbles show orange CL." To test this "null hypothesis" we then examine a sample population of Carrara marbles. If one sample is found not to be orange then the null hypothesis is found to be false. A key aspect of scientific methodology is that a hypothesis must be falsifiable. If the null hypothesis is incorrect – not necessarily "untrue" – then maybe an "alternative" hypothesis is better, such as the statement "MOST Carrara marbles show orange CL". Another way of looking at this is to say that "there is a high probability that a sample of Carrara marble will show orange CL" and that this *mode* will be observed in almost all cases. This second hypothesis is *confirmed*.

8.3.2
Probability

Probability is conceptually central to inferential statistics, but not for descriptive statistics. Probability is the relative frequency with which an event occurs. It is also possible to discuss distributions and many of their descriptive parameters without recourse to a discussion of probability as well.

Classical probability for independent events with equal outcomes: this rule states that for independent events that are equally probable, where an event does not in any way determine the occurrence of another event, such as say height and eye color, then by knowing the total number of possible outcomes we can calculate the probability for a particular outcome or event. If f is the observed outcomes or events and N is the total number of possible outcomes or events then the probability P(E) is:

$$P(E) = f/N$$

The classic example for this is the toss of a coin with the "head or tail" outcomes so that P(E) is:

1/2 or 0.5

One notes that since the two possible outcomes, head or tail, cannot happen at the same time they are termed mutually exclusive. Independent and mutually exclusive outcomes are sometimes confused. They are not the same.

Another illustration of rule 1 would be the presence of 7 basalt stone tools in a sample assemblage of 40 tools where P (E) for this occurrence is 7/40 or 0.175.

By using a rule termed 'complementation', the probability of a nonbasalt tool [P (not E)] is simply:

$$P(\text{not } E) = 1 - P(E) \quad \text{or} \quad 1 - 0.175 = 0.825$$

These rules are central in the calculation of expected probabilities used to determine the chi-square statistic as we shall see.

Other fundamental laws of probability are:

1. A probability of zero means the event cannot happen.
2. A probability of one means the event must happen.
3. All probabilities must be between one and zero.
4. The sum of the probabilities of all simple events must equal one.

Probability distributions are models of a specific kind of random process. The key to any discussion of probability is the assumption of randomness or that the occurrence of a thing or event is unpredictable. Returning to the coin toss, we know the outcomes must be head or tail, but we cannot predict which outcome will occur on any particular toss. Probability estimates quantify uncertainty. By rule (3) above, the uncertainty lies between one and zero with an increased certainty (or concomitant reduction of our uncertainty) if the value lies closer to one than zero.

8.3.3
Types of Error

In statistical inference a decision to reject the null hypothesis when it is really true is called a type I error. Likewise, the acceptance of the null hypothesis when it is false is called type II error. The probability of making a type I error defines the *significance level* or α. The probability of making a type II error is denoted as β. To illustrate this we can suppose that the rejection region for a hypothesis is at the 2 standard deviation level or has a probability of 0.9544. If our estimate – mean – falls within the rejection region and it is really a member of that population, then we have a type I error. Cannon (2001:186) illustrates this with regard to zooarchaeological assemblages. In his example, if one concludes there is a trend in relative abundance (of species) in sample assemblages when in fact there is not, then this is type I error.

8.3.4
Confidence Interval and the Z-Score

To use statistics in the tests of hypotheses, we can examine a set of sample values and determine their mean and compare that to an expected value or a hypothesized value. Generally, we have an idea of the "population" mean so we can assess our samples against this parameter. In addition, the use of mean implies a range of values for the population. Anything outside this range, typically expressed in terms of the standard deviation, is considered in the rejection region of the population. We name the range of values in our population as the *confidence interval.*

The confidence interval, or CI can be observed or calculated as we shall see in our discussion of the Student's t-distribution below. The CI is simply an interval of numbers about a mean and a *confidence estimate* is a measure of how sure or confident we are that an observed value will lie within that interval.

One way of expressing the distance of an observational measurement, x, away from the mean is the Z-statistic or Z-score. The Z-score is calculated as follows:

$$Z = (x - \bar{x})/\sigma$$

For example, a $\bar{x} = 728$ mm and 73 observations, x = 801, we define as a new statistic, the standard error of the mean, given by

$$\sigma_{\bar{x}} = S/\sqrt{N}$$

Z = 1 means the value of x is exactly 1 standard error from the mean and will occur, in the one-tailed case, 34% of the time. Here, again, it is best to use the parameter, $s_{\bar{x}}$ for the sample distribution's standard error rather than that of the population or instead of N we choose $n - 1$ for the denominator and derive the square root. The standard error of the mean observed from the samples drawn from the sample population. To calculate the Z-score, one *must* use sigma (σ) or the population parameter.

For our example:

$$Z = (801 - 728)/73 = 73/73 \quad \text{or} \quad Z = 1$$

One can use the Z-score to determine the probability of finding a specific value in a normally distributed population. One must use a table of Z-scores to convert to curve area. For Z = 1 this is self-evident, but for fractional values a table of Z-scores is necessary. This technique is introduced for inference more to illustrate the concept of standard error than to imply it is commonly used. Because the Σ is so rarely known the Z-score is of limited value. Statisticians such as W.S. Gossett developed the Student's t-distribution and the *t* test in 1908, which are more commonly used than Z. The t-distribution has a similarity to the normal curve and as sample sizes become large enough, closely approximates it.

The t-score, determined by the *t* test, differs from the Z-score in that the standard deviation of the population must be estimated. This is not surprising as most population parameters are not well known at first if ever. The confidence interval, CI, used in the t-distribution is calculated as follows:

$$CI = \bar{x} + (t_{0.05,n-1})(s_{\bar{x}})$$

We can intuitively estimate the age range of an archaeological site based on ceramics or we can estimate the range in the size of ceramic vessels of a particular phase or period. From that intuition we can then estimate the standard deviation to a first approximation. After more rigorous study of a large suite of samples that estimate can be refined such that the population's variability is more correctly known.

In calculating the t-score, the value is very much dependent on sample size because the standard error is that of the sample, s, where n is now $n - 1$. The calculation is similar to that of the Z-score. The direct calculation of t is as follows:

$$t = (\bar{x} - \mu)/s_{x_-}$$

This score can be compared to tabular values of t to examine whether a mean lies within a specific confidence interval. For example if a t-score of 2.3 is calculated, it exceeds the tabulated t-score of 1.984 so this mean is unlikely to be within the sample population's range or exceeds the significance level for the test.

Another measure of variation one can use, as it is expressed as a readily identifiable percent value, is the Coefficient of Variation (CV). It is calculated using the standard deviation and the mean:

$$CV = 100s/\bar{x}$$

Note that this parameter is unadjusted for the sample size.

8.3.5
Sample Size

More often the question facing a researcher is how large a number of observations is sufficient to draw a valid statistical conclusion as to a population's underlying characteristics. How many coins must we examine with microprobe or PIXE analysis before we can make valid inferences as to their composition as a group? How different this can be since by inspection of formulae like that for the standard error of the mean has n clearly defined, e.g., $S_x = \sigma/\sqrt{n}$. If we know the standard error and the standard deviation, then n is immediate. Generally, we do not know the parameters for a population we wish to study.

An important element in the discussion of how large a sample, n, is necessary for valid estimates of population mean follows from considerations of the probability of that sample estimate being reliable for the population. A basic definition of the probability of a mean is:

$$P_r = \lim X^n/N$$
$$N \to \infty \ \text{j}$$

As the limit approaches some N, the value of P_r will be equal to 1 (Johnson and Leone 1964). It follows from this that as n increases so will the probability that the sample mean will approximate that of the population. This is the Law of Large Numbers. Another important factor in choosing sample size is the homogeneity or heterogeneity of the population. Homogeneous populations require smaller samples and heterogeneous ones require larger samples. What is the difference between two means? If the difference is large, then the estimate is "crude" and requires fewer observations. If the difference is a fine one, then n will increase accordingly. There is some help in the concept of detectable difference as one does not need to know σ precisely, only the ratio between the difference, S, and σ (Sokal and Rohlf 1981). For purposes of discussion let us examine an example using *Homo erectus* skulls.

8.3.6
Example 1: Cranial Capacity in *H. erectus*

Given the following cranial capacity parameters: $\bar{x} = 935\,\text{cc}$; $\sigma = 132\,\text{cc}$ for *Homo erectus* samples, how large a sample would be required to be 90% confident that they would average greater than $920\,\text{cm}^3$ in cranial capacity?

Chebychev's Rule for Samples and Populations. The Russian mathematician, P.L. Chebychev formalized the rule that holds for any data set. The first property of this rule states that 75% of all observations will lie within two standard deviations of the mean; the second property states that ~90% will lie within three standard deviations on both sides of the mean (Weiss and Hassett 1993:98). Recalling our discussion of the Z-score, its meaning now should be more apparent. Chebychev's rule gives us the means to constrain the variability inherent in our samples. For example, the third property of this rule states that for any number, k, greater than 1, observations in a data set, at least $1-1/k2$ of the observations must lie within k standard deviations of the mean.

In this problem, we set the level of probability, e.g., how probable is our estimate. This done, the sample size determines precision and not population size (Kish 1965). This is echoed by Lazerwitz (1968) in the statement that population size has nothing to do with the size of the probability sample from that population. Certainly, as Lazerwitz points out, the relative homogeneity of the population will influence the sample size.

The central limit theorem (see Thomas 1976) states that if "random samples are repeatedly drawn from a population with a finite mean μ and variance σ^2, the sampling distribution of the standardized sample means will be normally distributed with $\mu_{\bar{x}} = \mu$ and the variances $\sigma^2 = \sigma^2/n$. The approximation becomes more accurate as n becomes larger.

Key to all this discussion of sample size is that for valid implementation of any of these formulae, one must draw the sample in a manner such that the assumptions of statistical theory are applicable (Cowgill 1964). This necessitates that the samples be drawn in a randomized manner.

With these assumptions and the new parameter $\sigma_{\bar{x}-}$, we can now calculate the sample size by suitable manipulation of the following two formulas:

(1) $Z = \bar{x} - \mu/\sigma_{\bar{x}-}$ and

(2) $\sigma_{\bar{x}-} = \sigma^2/\sqrt{n}$

From the above problem, we substitute the following values into (1): $935\,cm^3 = \mu$, $920 = \bar{x}$, and $\sigma = 132\,cm^3$. Solving for $\sigma_{\bar{x}-}$,

$$\sigma_{\bar{x}} = \frac{920 - 935}{1.29}$$

where $Z = 1.29$ and is determined by our level of confidence, 0.10. Hence, $\sigma_{\bar{x}} = 11.627$ and by formula (2), after rearrangement, or

$$n = (132/11.627)^2$$
$$n = 129 \text{ skulls}$$

Having reached this value for n, it is equally valid to ask if sample size of 129 *H. erectus* skulls is really realistic given the rarity of fossil remains of this hominid? It probably isn't.

This determination of sample size assumes σ is known. If σ is not known, then a preliminary sample is necessary to estimate this parameter. Past surveys can aid in this problem. Kish (1965) has suggested that if σ is not known we can guess CV, the coefficient of variation, where $CV = \sigma/\bar{x}$. CV is far less variable, hence with an estimate of CV and $\bar{x}$ we can "guesstimate" σ.

Another more common method is suggested by Thomas (1976):

$$\mu = np$$

where n = no. of trials and p = the probability of occurrence and, from the rule of complementation, q $= 1 - $ p, implying that

$$\sigma = \sqrt{npq}$$

This approach assumes that some notion of probabilities for events is known.

When p and q are not known, Lazerwitz (1968) utilizes the fact that the numerator in the formula for the variance of a sample proportion reaches its maximum value when the proportion is 0.5. Utilizing this value, a conservative estimate for sample size by the following formula:

$$n = \frac{(2)^2_{pq}}{[2SE(p)]^2}$$

where SE(p) = $\sqrt{pq/n}$.

The general formula is:

$$n = 1/k^2$$

where k is the desired interval about 0.5 at the 95% confidence level (cf. Chebychev's rule).

8.4
Data Analysis

8.4.1
Example 2: Single Variable Data – Phosphorus in Soils

In this example, observations are made on the presence of P_2O_5 in the sediment samples. From Chap. 7 we have seen that the analysis of labile or bound phosphorus is an important chemical indicator of archaeological sites. In 9 samples(n) we observe a range for phosphorus from 192 to 228 µg/g. The calculated mean is 218 µg/g P_2O_5 in the sediments. The variance, s^2, is 124.36 and the standard deviation, s, equals 11.1 µg/g. The standard error of the mean, $s_{\bar{x}}$, is 3.7 µg/g.

What is the variability? If we use the coefficient of variation, it is:

$$CV = (11.1)(100)/218 \text{ or } 5\%$$

a somewhat "large" variation. Using the t-score, we can establish a confidence interval such that 95% of the time for our sample mean will lie within it. From $t = \bar{x} - \mu/s_{x-}$, at 8 degrees of freedom (9–1), $t_{0.05} = 2.306$, and $s_{-,}\bar{x} = 3.7$ µg/g. Therefore, our 95% CI = 218 ± (2.306)(3.7) which means that the $\bar{x} = 218$ will be found within the interval of 209.51–226.51 μg/g. Any phosphorus value outside this range is outside the sample population and its (assumed) normally distributed variation. The value of 228 μg/g in our data is outside this population.

8.4.2
Data Analysis of More Than One Variable

The eminent statistician, John Tukey, enjoins us to plot the data . . ."there is no excuse for failing to plot and look" (1977). This should be the first step in data analysis. In plotting single variable data, we look for the central tendency in the distribution of the occurrences of that variable in populations. Our measures of dispersion and tests for fit or membership within a sampled population validate (or invalidate) the behavior of that variable. For instance, if the amount of bound phosphorus is increasing in an archaeological site, then we can speculate on the intensity of the occupation at that location. If we take our inquiry a step further, we may wish to examine the variation in the phosphorus amount relative to a second variable such as either time or space. Pollen studies of sediments of archaeological interest reveal variations in individual plant species, arboreal or nonarboreal. These variations in particular plant species are typically plotted in so-called pollen diagrams. In these familiar bivariate plots the pollen amounts for a species are compared to their distribution across time. If we assign the variable name, y, to the pollen amount of a given plant species then, in mathematical parlance, it can be viewed as functionally related to a second variable, x, which in the example here, is time. Written as a function of time (x), pollen intensity is:

$$y = f(x)$$

8.4.3
Covariation-Correlation, Causality or "Not"

The basic question for most studies of relationships in two or more variables is – "Does x cause variation in y?" If not, there is no functional relationship between x and y. Even if there is covariation in two variables, there may not be a causal relationship. Correlation does not imply causality. In the study of causality we try to discover the underlying conditions that exist between x and y such that (1) y may never occur without x, a logically *necessary* condition or (2) x is always followed by y of which it is "a" cause, but necessarily the only cause. This second situation is termed a logically *sufficient* condition. It is possible to establish correlation with statistics, but not causality.

In statistics as in other experimentally based inquiry, experimental design provides greater certainty and certainly, increased efficiency in testing relationships in x and y. Good experimental design attempts to formulate exploratory studies that will hopefully, isolate variables which do exist in some causal relationship and then design tests that accurately portray the frequency of occurrence that supposed relationship.

Implicit in the determination of the frequency of the concomitant occurrence of x and y is an increased reliability in the suspected causation. Another aspect of good experimental design is the accuracy of measurements of x and y. Repeatable replicates (precision) of accurate measures of x and y increase the reliability of observed correlations and reduces or minimizes bias in an experiment. Bias is related to error, systematic or otherwise, and is generally related to measurement in statistical tests (Kempthorne and Allamaras 1986). One can express bias as the residual errors in a data set as that of the root-mean-square (RMS) error of an estimate (mean, average).

8.4.4
Least-Squares Analysis and Linear Regression

The simplest relationship between more than one variable is bivariate-x and y. The simplest relationship between x and y is linear such that:

$$y = a + bx$$

e.g., the slope-intercept equation for a line. The general linear model for n the regression of one variable y on another variable x, which exist in direct variation, is:

$$y = \beta x$$

where β is the population parameter for slope and whose range is $-1 \leq \beta \leq 1$. In a perfect correlation between x and y, $y = x$ and $\beta = \pm 1$. The perfect negative correlation is the case of minus (-1). If there is no correlation, $y \neq x$ and $\beta = 0$.

In common practice, the correlation coefficient is written as r and is equal to $r = b_{y \cdot x} \cdot s_x / s_y$.

Here, b is the slope of the regression of x and y and this is multiplied by the ratio of the standard deviations of x and y.

One can graphically, examine correlation in the regression of a least-squares plot (Fig. 8.3). (1) is a case of a perfect correlation in x and y; (2) a weak correlation and (3) no correlation, e.g., y = z no matter the value of x. The least squares method is so named for the procedure's criterion on making the sum of squares for deviations (Δy), from the regression line, as small as possible (M. Penita, pers. comm.). Another way of stating this is that the mean square $(\Delta y)^2$ difference between the data and the straight (regression) line should be a minimum. A brief explanation of this is as follows. The equation $y = a + bx$, based on two values of x and y clearly define a line, but when we introduce a new, second, value for x, simply called, x_1, the solution is not y, but y_1 which is now equal to $a + bx_1$ and differs from the line $y = a + bx$ by Δy. The

Fig. 8.3. Example of a typical regression line-of-best fit through a bivariate (x,y) data set. (Adapted from Waltham 1994, Fig. 7.6, p. 120)

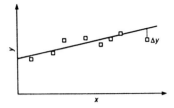

A straight line drawn through the x–y data such that it passes close to all of the points. The deviation, Δy, of the line from the last point is also shown. The deviation could be found for each of the other seven points as well. The best fit straight line is that which produces the smallest possible $(\Delta y)^2$ averaged over all of the points.

process for fitting a line to a series of x_n, y_n pairs is called linear regression. From this procedure we can evaluate the correlation coefficient, r.

Two ways of calculating r involve (1) calculation of b, the slope $(y_2 - y_1)/(x_2 - x_1)$ or (2) a direct method of using x and y. The first, (1), is:

$$b = \left[\Sigma(x - \bar{x})(y - \bar{y})\right] \big/ \Sigma(x - \bar{x})^2$$

and

$$r = b \cdot s_{\bar{x}} \big/ s_{\bar{y}}$$

where $s_{\bar{x}}$ and $s_{\bar{y}}$ are standard errors of the means for x and y. The second, (2), is more direct:

$$r = \Sigma xy \big/ \sqrt{\Sigma x^2 \Sigma y^2}$$

Most regression lines do not cross the y-axis at zero so the y-intercept value has implications for assessing a data set. For instance, if the y-intercept for a set of fission track counts vs. time is found to be greater than time (t) = 0 (y-axis) and is rather, for purposes of demonstration, equal to 1550 years, then the regression line is not appropriate for samples less than this age.

The correlation coefficient, r^2, is a ratio of the variance – that of the original variance in y to that of the sum of squares of residuals or deviations, Δy, from the regression line of best fit (Fig. 8.3). Values of r^2 could be 1 or 0 with the values somewhere in between. A high correlation coefficient means a good fit of the regression line to the observed data. For example:

$r_2 = 0.78$, a high, positive correlation in x and y;

or

$r_2 = 0.25$, weak correlation.

Hawkes (1995) lists the following descriptions for the correlation coefficient.

$-r = 0$	no linear relationship
$-0.5 < r < 0.5$	weak negative linear relationship
$0 < r < 0.5$	weak positive linear relationship
$-0.8 < r < -0.5$	moderate negative linear relationship
$0.5 < r < 0.8$	moderate positive linear relationship
$-1.0 < r < -0.8$	strong negative linear relationship
$0.8 < r < 1.0$	strong positive linear relationship
$r = 1$	exact positive linear relationship
$r = -1$	exact negative linear relationship

8.4.5
Example 3: Alloys in Iron Age Metallurgy

An example of least-squares analysis of data from a rather straightforward PIXE study of the compositional makeup for a suite of Iron Age (800–58 B.C.) metal artifacts to ascertain their nature as to either: (1) copper, (2) bronze or (3) brass. The difference being based on the occurrence of tin (Sn) in less than trace amounts in bronze; the occurrence of zinc (Zn) in similar compositional amounts (~10%) in brass while only copper (Cu) predominates at levels of 97–100% in "pure" copper artifacts. In the data from Table 8.2, one observes

Table 8.2. Compositional data (%) for Iron Age artifacts

Artifact		Sn	Pb	As	Sb	Ag	Ni	Bi	Co	Zn	Fe	Cu
170	2P	9.87	0.041	0.085	0.052	0.012	0.069	0.001	0.035	0.022	0.064	89.7
171	3P	6.13	0.017	0.165	0.060	–	0.187	–	0.027	0.014	0.027	93.4
172	3N3	5.23	0.198	0.182	0.105	0.019	0.56	0.002	0.033	0.015	0.094	93.6
173	6N2	9.69	0.044	0.34	0.62	0.020	0.25	–	0.014	0.008	0.007	89.0
174	5N1	9.00	0.27	0.108	0.193	0.094	0.137	0.011	0.024	0.018	0.030	90.1
175	1P	9.42	0.61	0.093	0.092	0.133	0.065	0.021	0.026	0.056	0.23	89.3
176	1P	0.135	9.19	0.032	0.018	0.140	0.004	0.033	0.007	0.55	0.77	89.1
177	5P	0.017	0.50	0.020	0.085	0.061	0.034	0.011	0.061	0.24	1.89	97.1
178	2P	0.121	0.47	0.32	0.007	0.075	0.071	0.031	0.060	0.64	1.02	97.2
179	1P	0.81	0.046	0.072	0.005	0.119	0.002	0.017	0.021	0.021	1.60	97.3
180	1P	9.69	0.49	0.090	0.074	0.099	0.041	0.021	0.034	0.060	0.60	88.8
181	3N2	10.70	0.30	0.139	0.075	0.085	0.26	0.019	0.025	0.030	0.43	87.9
182	6P	0.050	0.41	0.030	0.077	0.056	0.030	0.012	0.053	0.174	0.90	98.2
183	3P	0.110	0.061	0.017	0.006	0.074	0.035	0.001	0.034	0.134	2.02	97.5
184	4P	0.016	0.57	0.018	0.020	0.097	0.042	0.021	0.042	0.24	1.35	97.6
185	2P	0.095	1.16	0.030	0.018	0.079	0.024	0.003	0.025	0.34	0.193	97.3
186	2P	0.083	0.047	0.054	0.007	0.044	0.020	0.003	0.069	0.029	0.98	98.7
187	3P	–	0.009	0.003	0.001	–	0.008	0.006	0.001	0.003	0.018	100.0
188	6N2	5.19	1.38	0.23	0.48	0.157	0.23	0.013	0.034	0.008	0.016	92.3
189	6P	11.21	0.41	0.083	0.119	0.086	0.075	0.016	0.021	0.019	0.031	87.9
190	6P	7.70	0.099	0.048	0.081	0.065	0.045	–	–	0.010	0.101	91.9

little variation in Zn, which is consistently at trace levels while the variation in Sn is such as to suspect a separation in these artifacts into bronze and copper artifact classes. Simple regression analysis is used for bivariate pairs of x, (in all three cases the value of Cu) and y (alternately Sn then Zn). The correlation coefficients for each paired regression were:

1. Cu vs. "high" Sn (greater than 5.19%) artifacts, r = 0.97, r^2 = 0.94.
2. Cu vs. "low" Sn (less than 5.19%) artifacts, r = 0.53, r^2 = 0.28.
3. Cu vs. Zn (n = 21) artifacts, r = 0.68, r^2 = 0.46.

Only in case 1. Cu (x) vs. "high" Sn amounts (y) is the correlation significant (r^2 = 0.94).

To determine the significance of the r value, in this case 0.97, a t test should be done on either the slope parameter, β, or on the r value itself. Either is easily done. For β, or b, the test is:

$$t = b - 0/s\beta$$

For a t test of r we use the following formula. We compute the standard error, $s_{y \cdot x}$, by:

$$s_{y \cdot x} = \sqrt{1 - r^2/n - 2}$$

where n is simply the number of x − y comparisons. The t test becomes:

$$t = (r - o)/\sqrt{1 - r^2/n - 2}$$

Note its similarity to the test using the slope parameter, β.

In regression studies there are two variations, termed model I and model II or Bartlett's method. They differ in how the variable x is handled. In model I regression, x is fixed and set as such in the experiment; in model II both x and y are random. Model I regression studies are common in psychological, education and agricultural testing. In most instances, and in particular when dealing with archaeological and geological phenomena, Model II regression is done. Thomas (1976) states that "r" is meaningless in model I regression. The correlation analysis assumes a bivariate normal distribution which is not the case when x is nonrandom and fixed. This is to say the statistics of r and r^2 are grounded on very different assumptions depending upon whether the data are collected using model I or model II (Thomas 1976). Many researchers today prefer to use what is termed Major Axis Regression Analysis rather than least squares regression (model I) recognizing that both x and y vary at random (Steve Holland, pers. comm.).

8.4.6
Paired Comparisons or Paired Difference Experiments

One can view paired difference comparisons as "before-and-after" types of tests. For instance, one can map the positions and/or the amounts of elements in artifacts for comparison before-and-after something has been done to the material, such as the weathering of a surface.

One can use a reference material for comparison as well. What the paired comparison test seeks to discover is the presence of any difference in the two samples and to quantify that difference (Hawkes 1995). One could use the sequence of two tree-ring samples for comparison. Certainly a comparison that yields no difference in two ring sequences implies a similar age for both. This is what is done in dendrochronology to achieve a "match" between two sample sequences.

The dendrochronologist, in this case, desires to find no difference in the match between the unknown and the master reference tree ring sequence . In paired comparisons of sequenced data, one can be looking for similarity or difference. An excellent example of both the type of data and the utility of the test can be found in recent studies of Neanderthal mtDNA sequences.

Recently (1997), the fossil remains of the original 1856 Neanderthal find was assayed using modern DNA analysis (Krings et al. 1997). A sample of the right humerus from the ancient hominid was used to obtain a 360 base pair (bp) sequence of mitochondrial DNA (mtDNA). This sequence was then compared to a modern human reference standard as well 994 contemporary human mitochondrial lineages (478 Africans, 494 Asians, 167 American Indians and 20 Australian/Oceanic individuals) together with 16 nonhuman, chimpanzee, lineages for a total of 2051 human and 59 common chimpanzee sequences. This study was done to address questions of the phylogenetic relationship of Neanderthals to modern human populations as well as the great apes. The null hypothesis would be something of the nature of no difference in the 360 bp Neanderthal mtDNA sequence and modern humans. A second null hypothesis might be no difference between Neanderthals and the great apes based on the DNA comparison.

In a pairwise comparison of the respective mtDNA sequences, some interesting results were obtained. As illustrated in Fig. 8.4, the observed differences in human–Neanderthal sequences averaged 25.6 ± 2.2 differences; between human–human sequences an average of 8.0 ± 3.0 differences and between human–chimpanzee an average of 55 ± 3.0 differences. Using a calculated rate for mutations to arise in the sequences, a common ancestor for humanity (including Neanderthal) existed up to a divergence at 4–5 Ma ago and a divergence in the human and Neanderthal lines around 550–690 ka ago. The researchers concluded on the basis of the pair-wise comparison that the mitochondrial DNA sequence for Neanderthal lies outside modern human variation. What the researchers do not report is a test of significance for their paired comparison results. This can be done using the t test. For paired comparisons the t-statistic is found as follows;

$$t = \bar{x}_D - \mu_D / S_D / \sqrt{n_D}$$

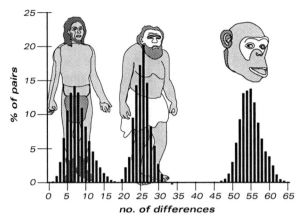

Fig. 8.4. Distributions of pair-wise sequence differences among humans, Neanderthals and chimpanzees. (Adapted from Krings et al. 1997, Fig. 6)

Here, we examine the sample differences, $\bar{x}_D$ using the standard deviation of the differences, S_D, where n should be apparent; the population mean, μ_D, is presumed to be zero.

8.5
Exploration – Multivariate Approaches to Large Data Sets

8.5.1
Exploring Correlation Among n Greater Than Two Variables – Some Multivariate Statistical Techniques

Before the advent of digital computing, multivariate or multidimensional data were difficult to analyze statistically due to the laborious nature of the calculations where, in sets of data of n greater than three variables, the use of matrix algebra was required. In the last decades of the twentieth century the use of vector-based multivariate analytic techniques has increased in almost all sciences. The most frequently encountered multivariate technique(s) are found under the general heading of factor analysis.

Factor analysis methods – principal components, Kaiser image, Harris image, iterated principal axis, to name but some of the more common – are described and discussed in detail elsewhere (Mulaik 1972; Gorsuch 1984; Davis 1986). In factor analyses, linear algebra is used to manipulate variables in a data set to examine possible relationships and associations among them. Each variable is treated as a vector within n-dimensional space where n is equal to the number of variables, e.g., 12 variables, 12 vectors. The data set of a matrix of n variables measured against n number of samples (cases).

Two patterns of relationships can be examined using factor analysis – R and Q-modes. The former, R-mode, factor analysis attempts to isolate relationships among the variables in the data set whereas Q-mode factor analysis attempts to examine relationships among the various samples or cases. On most archaeological or geological studies Q-mode is the factoring technique most used. The end result of either approach is a series of "factors" or members that determine the commonality estimates among the variables. The key characteristic of a factor analysis is that, if done correctly, it will reduce the number of "variables" needed to describe the data. The new "variables" are called factors and are linear combinations of the original variables. Generally, in factor analyses the rule of thumb is that the maximum number of factors is no more than one-half the number of variables being analyzed. In practice the number of factors are well below half and number about four to six in even the largest data sets. The objective is to account for as much of the variance as possible in the original data set. Factor analysis algorithms operate on matrix data that are sets of linear operations of either the raw data (variables). The basic linear equation is:

$$S_n = f_1c_1 + f_2c_2...f_nc_n + R$$

S_n is the bulk amount of the variables, 1 to n, in a sample; f is the fraction of the factors, 1 to n, in the sample; C is the amounts of the variable in the factors; and R is the residual terms for error. For eight variables there will be eight linear equations. The mathematical strategy involved is the determination of a best estimate for values of factors extracted by minimizing the values for the residual errors in R. The method most often used to do this is the now familiar method of least-squares.

A principal assumption for justifying the use of least-squares is that the data are normally distributed. The least-squares approach minimizes the sum of squares of the residuals, R. A cutoff value for the final number of factors can be done in several ways. The most common is to choose the number of factors that account for at least 95% of the total variance. It is common for the first factor to account for nearly 50% or more of the variance, with the other factors accounting for less and less. A rule of thumb is that as the amount of variance accounted for by a factor decreases, the variance of its components increases. In the principal components method of factor extraction (PCA), eigenvalues (and eigenvectors) are used as the second way of factor selection. An eigenvalue can be viewed as a characteristic root of the matrix. Because of the n-number of equations there are n-eigenvalues, but not all are important in accounting for the total variance of the data set.

In general, the number of significant factors extracted by a factor analysis is equal to the number of eigenvalues determined: One rule of thumb in the use of eigenvalues is that the variance accounted for must be at least 75%. Another selection criterion for factors is the eigenvalue be greater than or equal to one. The maximum or largest eigenvalue is equal to the variance of y. Therefore, for a matrix of r (nonzero) eigenvalues there are exactly r principal components.

8.5.2
Discriminant Function Analysis and Principal Components Analysis

Discriminant (function) analysis (DA) presumes that a set of variables, such as elemental data, can distinguish (*discriminate*) between groups, for example pottery clay sources that contain these elements. DA improves the distinction between sources by statistically removing variables (elements in this case) that have any or no discrimination value (Klecka 1975). Discriminant analysis works best between a priori defined groups.

In discriminant function analysis the "distance" – near or far between two groups (minerals, artifacts, etc.) in a statistical sense is compared by use of what is termed *pair-wise generalized squared distance function,* D^2. The main idea behind discriminant function analysis is to define a function, Y, which will discriminate between groups of p variables. For the best discrimination the respective group means, of say two groups Y_1 and Y_2, should be widely separated.

An analytical software program (SPSS) calculates the coefficient vector and the distance function, then projects the groups onto a *linear discriminant function, Z,* defined by variables within the groups (e.g., SiO_2, Pb, Mg, etc.). Using the value D^2 (E), which some readers will recognize as the Mahalanobis distance, a *posterior probability,* P_r, for a sample (E) belonging to different groups. In general, membership of a sample (E), in a group is indicated by a high P_r, say, 0.8–1.0. A low P_r, e.g., less than 0.5 or so indicates nonmembership in a group.

In principal components analysis, or PCA, the factors extracted from a correlation matrix are presented as axes. The first principal component or axis accounts for the maximum possible variance; the second the maximum remaining variance and so one can use this axis extraction for two-axes plots wherein the number of samples are displayed as a scatter plot. Clusters can be seen in many of these plots that infer relationships between the samples based on the derived and measured distance from centroids of the clustered samples. Figure 8.5 illustrates the use of PCA to isolate elemental groupings in marbles from Mount Pentelikon, Greece. In Fig. 8.5 we see a bivariate plot of the first two principal components illustrating correlation of elements in the marbles. Closely spaced elements indicate strong correlations. Another example involves the examination of the elemental composition of prehistoric stone from the Great Plains of the United States. Table 8.3 lists the seven elements used in the principal components analysis (Hoard et al. 1992). The analysis was performed using the logarithms of the elemental concentrations. This is not an uncommon practice since otherwise nontransformed values for the major elements could dominate the analysis (Kallithrakas-Kontos et al. 1993). The plot (Fig. 8.6) of the two principal components shows the sim-

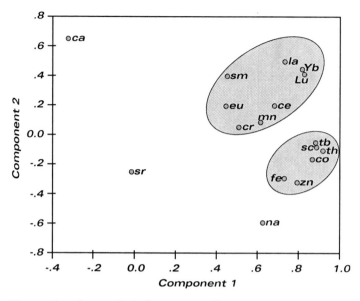

Fig. 8.5. Plot of two principal components for elements found in Pentelic marble samples. (Courtesy of Scott Pike)

ilarity of the Colorado materials and their relative distinctness from the South Dakota materials. One sees a slight overlap between the two groups, but they are clearly distinct. When the 12 ceramic samples from a Kansas site (Eckles) are included in the plots of the source groups, it is possible to decide if either is a candidate as a place of origin (Hoard et al. 1992). In 10 of 12 cases, the Eckles site lithic materials fall within the Colorado source.

An important question raised by Baxter and Jackson (2001) concerns variable selection in artifact compositional studies. These authors point out that

Table 8.3. Principal components of seven elements used in the study of Great Plains prehistoric stone. (Adapted from Hoard et al. 1992)

Variation explained (%)	PC1 48.4	PC2 26.4	PC3 12.8	PC4 6.6	PC5 4.7	PC6 1.0	PC7 0.1
As	−0.28	−0.08	−0.95	0.04	−0.20	0.01	−0.01
Sm	0.26	0.46	−0.15	−0.33	0.27	−0.06	0.71
U	0.32	0.44	−0.15	0.32	0.31	0.06	0.71
Yb	−0.85	0.44	0.20	−0.16	0.11	−0.01	−0.04
Cr	0.12	0.56	−0.04	0.81	−0.10	−0.01	−0.01
Zn	0.08	0.28	−0.03	−0.30	−0.87	0.24	0.00
Al	0.02	0.06	−0.02	−0.09	−0.21	−0.97	−0.09

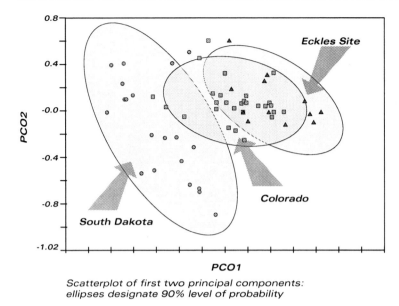

Scatterplot of first two principal components:
ellipses designate 90% level of probability

Fig. 8.6. Two principal components of Great Plains prehistoric ceramics. (Adapted from Stewart)

compositional studies such as those just described proceed from a first principle used in biological numerical taxonomy – "the greater the content of information in the taxa of a classification and the more characters on which it is based, the better the classification will be" (Sneath and Sokal 1973). Because instrumental methods, such as NAA and the ICP variants (cf. Chap. 7), routinely determine 30–40 elements per sample, it would appear that they meet the foregoing injunction. Harbottle (1976) affirmed this approach in his influential paper. Baxter and Jackson take issue with this approach suggesting instead that 'less is more' with regard to the number of analytical variables used in certain analyses (Baxter and Jackson 2001). Their point, in the context of multi-element-multivariate analyses, is that too often the choice of many variables may obscure the perception of actual patterns (Fowkles et al. 1988; Kzanowski 1988; Baxter and Jackson 2001). Kzanowski illustrates his point by pointing out that sample size may militate against the use of too many variables, particularly where in cluster or factor extraction the number of cases used in the analysis must exceed the number of variables used. Instances where this injunction has been ignored extend back to the early days of the use of multivariate statistics in archaeology in general, and in archaeometry, in particular. In these cases some form of variable reduction seems mandated along with the use of PCA.

The use of the discriminant function as well as principal components analysis has a long history in attempts to classify archaeological materials

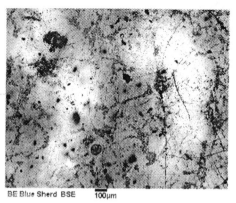

BE Blue Sherd BSE   100µm

Fig. 8.7. Spode blue printed design *left*; backscattered electron (BSE) image of dye. (Douglas 2000).

(Sieveking et al. 1972; Ward 1974; Picon et al. 1975; de Bruin et al. 1976; Leach and Manley 1982; Craddock et al. 1983; Rapp et al. 1984; Vitali and Franklin 1986). In particular, the discriminant function mathematically generalizes the characteristics of a group of objects and can be used to evaluate the probability of membership of any object of a group (Vitali and Franklin 1986). The underlying assumption in these types of provenance analyses is that the elemental composition of pottery is characteristic of its clay and thus of a particular site or place of manufacture.

Douglas (2000) has extended this type of analysis to historic English ceramics in a study of pigmenting elements in the dyes used for the printed designs found on nineteenth-century pottery. This study concentrated on 16 samples of Spode pottery manufactured at Stoke-on-Trent, England. A JEOL 8600 Superprobe (EMP) was used in all the instrumental analysis of the underglaze dyes (Fig. 8.7). Table 8.4 lists Douglas' use of the discriminant function on these 12 ceramic samples based on SiO_2 and PbO content.

The analysis clearly shows two distinct groups of "Age 0" and "Age 1" which stand for Spode pottery manufactured before 1834 (Age 0) and after 1834 (Age 1). Like the analysis of the Great Plains prehistoric samples, a Spode pottery sample in this case, of unknown chronological provenance, can be examined instrumentally and classified with some confidence.

8.5.3
Cluster Analysis

A variant of multivariate analysis is termed average-link cluster analysis and it seeks to measure association in the samples, but unlike principal components analysis it provides little information about the variables themselves.

Table 8.4. Spode pottery analysis using linear function. (Douglas 2000)

Obs	From age	Classified into age	Age 0	Age 1
BS-1	1	0	0.78	0.22
BS-2	1	1	0.0354	0.9646
BS-3	1	1	0.1627	0.8373
BS-4	1	1	0.0073	0.9927
BS-5	1	1	0.0003	0.9927
BS-6	1	1	0.0034	0.9966
BS-7	1	1	0.0018	0.9982
BS-8	1	1	0.0002	0.9998
BS-10	1	1	0.0041	0.9959
BS-11	0	0	0.7305	0.2695
BS-13	0	0	0.9185	0.0815
BS-14	0	0	0.9784	0.0216
BS-15	0	0	0.9849	0.0151
BS-16	0	0	0.9998	0.0002
BS-17	0	0	0.9998	0.0002
BS-18	0	0	0.9996	0.0004

Cluster analysis starts with determining similarity (distance) coefficients computed between pairs of samples rather than variables. Papageorgiou et al. (2001) point out a "problem" with methods of cluster analysis in that there are so many of them. Cluster analysis is the most widely used multivariate technique in archaeology (Papageorgiou et al. 2001). As pointed out earlier in this chapter, the output of instrumental analyses of artifactual and geoarchaeological materials produce extensive bodies of measurement data across many variables. Since most of these studies are exploratory in nature, a family of statistical methods such as those of cluster analysis will continue to enjoy popularity.

A common form of representing the output for cluster analysis is the dendrogram shown in Fig. 8.8. The dendrogram presents the samples ranked in association with those most similar. If the distance or similarity coefficients are printed for each set, then those values most different (ex. 1.15 and 0.092) mark boundaries between clusters. An interesting use of cluster analysis on data from instrumental analysis – XRF and PIXE was done by Kuhn and Sempowski (2001). In this study of artifactual material from early members of the Iroqouis League, ca. 1590, the Mohawk and Seneca tribes were examined as proxy data for increasingly strong political ties – a league – between these widely separated Great Lakes cultures. A cluster analysis of elemental means (13 elements) was done for each archaeological case (site) to identify natural groupings in the data and to indicate whether the site samples clustered into Seneca-Mohawk groups. The output of these analyses in

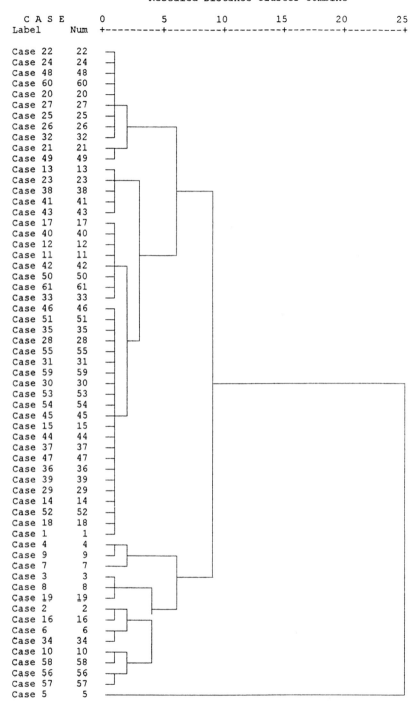

Fig. 8.8. Typical dendrogram plot of results of average-link cluster analysis of marbles from Carrara (IT) quarries of antiquity. Similarity is scaled from *left* to *right* with the samples most closely related between 0 to 5 on this re-scaled distance axis. (Herz et al 1999)

dendrogram form were similar to that in Fig. 8.8. Kuhn and Sempowski's results support a coalescence of the powerful league at the end of the sixteenth-century or the early seventeenth-century. Kuhn and Sempowski utilized principal components and discriminant function analysis as well.

In all cases of factor analysis, it is often difficult to determine precisely what the new grouping of variables or cases means. In the analysis of ancient coins the separation of key elements – noble vs. base metals – can enable the investigator to assign, with some confidence, identities to the factors separated on the basis of these metallic elements. In ceramics, the compositional variability of differing wares can often be used to isolate distinctness or similarity. This is particularly useful when one routinely can obtain reliable estimates for 30 or more elements using ICP or NAA making factor analysis methods essential for discriminating association among the key combinations of variables represented by the factors. In their study of prehistoric pottery clays with AAS (cf. Chap. 7), Singleton et al. (1994) utilized both cluster analysis and DA. Their results, first using DA to discriminate key chemical elements, led the researchers to refine their analysis to four elements that defined the spatial variability of the pottery.

8.5.4
Chi-Square: "Nonparametric" Calculation of Association

Not all data require comparison using the normal distribution and its associated parameters – mean, standard deviation, etc. Rather than using "parameters" we can choose to use the observed proportions in our sample to measure association by means of the chi-square statistic. The chi-square statistic and the chi-square distribution allow us to use the observations within the pre-determined categories – vessel types, axe varieties, weapons, sites, etc. – to make tests of association, independence and significance. The chi-square statistic basically measures departures from the average (Drennan 1996).

Categories are often called nominal data. They allow us to simply enumerate observations within them – 87 hand axes, 13 Solutrean points, 23 burins, 47 dolmens, etc. If the categories are arranged in some rank order, then we are dealing with the next scale of data – *rank*. Categories contrast to *measurements* made on interval or ratio scales. These measurements are quantified and used in the parametric tests we have discussed earlier. In the use of the chi-square, we assume no order or rank and we simply examine the proportion of the observations within categories such as those of types. Types cannot be ranked, whereas categories of size can. Still we count the representative sherds of types we recover and arrive at their numerical proportions. These values can then be used in the calculation of the chi-square statistic.

The chi-square statistic is calculated as follows:

$$\chi^2 = \Sigma(O_i - E_i)^2 \big/ E_i$$

where O_i is the observed number in any category; and E_i is the expected number in any category.

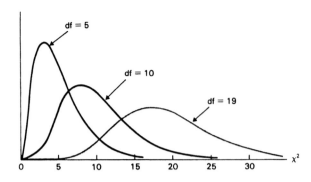

Fig. 8.9. Three chi-square distribution curves. Note the higher degrees of freedom (df = 19) approximates the normal distribution

The number of items in a category are counted and then compared to those of another category by use of χ^2. This value is then compared to some chi-square distribution (Fig. 8.9). By way of example, the simplest chi-square comparison is that of the 2×2 chi-square table.

In Table 8.5 are listed the number of observations of, say, the occurrence of Bronze Age metal axe types – winged and socketed – in two levels of a site. We ask the question whether there is a measurable difference in the occurrence of the axe types found in these two excavation levels. The table is constructed thus:

The observed values are apparent. To arrive at the expected values, E_i, for winged axes in level A, required by the chi-square formula, one multiplies the row total, 50, by the column total, for winged axes, 40 and divides this value by the grand total 80. Each "cell" of this table is calculated in a similar manner. These individual cell values are then summed as follows:

$$\chi^2 = (20-25)^2 \big/ 25 + (30-25)^2 \big/ 25 + (20-15)^2 \big/ 15 + (10-15)^2 \big/ 15$$
$$\chi^2 = 1+1+1.66+1.66$$
$$\chi^2 = 5.33$$

Comparing this value to tabulated chi-square values gives a measure of the similarity or difference in these categories from level A to level B. The number of degrees of freedom for a table is the product of one less the number of columns and one less the number of columns. Since the number of rows and columns are equal to two the calculation is $1 \times 1 = 1$ df.

At one degree of freedom, $\chi^2 = 5.33$ and 5.33 lies between the tabulated values of 3.841 and 5.412. These values represent significance at the 0.05 level (3.841) and the 0.02 level (5.412). Our measured value is closer to that of the 0.02 level so the similarity is significant at a very high level of confidence. In this result, it is highly plausible that the materials are contemporaneous and differ from the other level. Like our multivariate examples, the statistical analysis has provided some quantitative measure of group (level, type, age, etc.) membership and difference.

Table 8.5. Contingency table for winged vs. socketed axes

	Winged axes	Socketed axes	Total
Level A	20	30	50
Level B	20	10	30
Total	40	40	80

Similar to the normal distribution, the area under the chi-square curve is equal to 1. In Fig. 8.9, one quickly observes that these curves are skewed to the left for smaller degrees of freedom. It is also apparent that as the number of degrees of freedom increases, the chi-square distribution approaches the shape of the normal distribution. Chi-square is used on large and small samples. In small samples, only a very strong difference/similarity is significant as in our example. In larger samples, weak or small differences are magnified. Recall our recent discussion of variable selection here as well. In terms of significance, the larger the sample size, exaggeration of small differences within categories is a possibility. As in all cases of the use of parametric and nonparametric statistical measures, one must always consider sample size.

To end this chapter, and overall, this discussion of techniques for archaeological geology, let me paraphrase Simon A. Levin, in his MacArthur Award Lecture of 1989 (Levin 1992), wherein he notes that "the simple statistical (or analytical) description of patterns is a starting point; but correlations are *no* substitute for (a) mechanistic understanding". Statistics and the other techniques discussed in the preceding chapters allow us to describe phenomena and patterning in archaeologically interesting phenomena across a wide range of scales. However, it is for the archaeological geologist to use these techniques to infer conclusions of archaeological consequence.

References

Adams AE, MacKensie WS, Guilford C (1997) Atlas of sedimentary rocks under the microscope. Addison Wesley Longman, Harlow

Agassiz L (1840) Etudes des glaciers. Proceedings of the Geological Society of London, vol 3, Geological Society, London

Agassiz L (1967) Studies on glaciers; preceded by the discourse of Neuchâtel. Hafner, New York

Ahler S, Kvanne K, Kvanne JA (2000) Summary report on the 2000 field investigations at Fort Clark State Historic Site, 32ME2, Mercer County, North Dakota. State Historical Society, Bismarck, ND

Aitken MJ (1974) Physics and archaeology, 2nd edn. Clarendon Press, Oxford

Akridge DG, Benoit PH (2001) Luminescence properties of chert and some archaeological applications. J Archaeol Sci 28:143–151

Allard GO, Whitney JA (1994) Environmental geology lab manual. Wm C Brown, Dubuque

Ambroz JA, Glasscock MD, Skinner CE (2001) Chemical differentiation of obsidian within the Glass Buttes complex. J Archaeol Sci 28:741–746

Andersen ST (1986) Paleoecological studies in terrestrial soils. In: Berglund BE (ed) Handbook of Holocene paleoecology and paleohydrology. Wiley, New York

Anderson DG, Schulderein J (1985) Prehistoric human ecology along the upper Savannah river: excavations at Ruckers Bottom, Abbeville and Bullard Site Groups. Report prepared for the National Park Service by Commonwealth Associates, Jackson, MI

Anderson T, Villet D, Serneels V (1999) La fabrication des meules en grès coquillier sur le site gallo-romain de Châbles-Les Saux (FR). Arch Suisse 22(4):182–189

Andrews HE, Besancon JR, Bolze CE, Dolan M, Kempter K, Reed RM, Riley CM, Thompson MD (1997a) Igneous rocks and volcanic hazards. In: Busch RM (ed) Laboratory manual in physical geology. Prentice Hall, Upper Saddle, NJ, pp 50–68

Andrews HE, Besancon JR, Gore PJW, Thompson MD (1997b) Sedimentary rocks, processes and environments. In: Busch RM (ed) Laboratory manual in physical geology. Prentice Hall, Upper Saddle River, NJ, pp 69–92

Arnold B, Money C (1978) Les amas des galets un village littoral d'Auvernier-Nord (Bronze final; lac de Neuchâtel): études géologique et archéologique. Bull Soc Neuchâtel Sci Nat 101:153–166

Arnold DE, Neff H, Bishop RL, Glaswick MD (1999) Testing interpretative assumptions of neutron activation analysis. In: Chilton ES (ed) Material meanings. The University of Utah Press, Salt Lake City

Arrhenius O (1931) Die Bodenanalyse im Dienst der Archäologie. Zeitschrift für Pflanzenährung, Düngung und Bodenkunde. B10:427–439

Aspinall A, Lynam LT (1970) The induced polarization instrument for the detection of near surface features. Prospezioni archaeologiche 5:67–76

Aspinall A, Crummet JG (1997) The electrical pseudosection. Archaeol Prospection 4:37–47

Bahn P, Renfrew C (1999) Archaeology: a brief introduction, 7th edn. Prentice-Hall, Upper Saddle River, NJ

Ball T, Gardner JS, Brotherton JD (1996) Identifying phytoliths produced by the inflorescence bracts of three species of wheat (*Tritcum monococcum* L, *T dicoccon* Schrank, and *T aestivum* L) using computer-assisted image and statistical analysis. J Archaeol Sci 23:619–632

Ballard RD (1976) The Discovery of the Titanic. Warner Books, New York

Ballard RD (1993) The lost ships of Guadacanal. Warner Books, New York

Ballard RD (1994) Explorations: my quest for adventure and discovery under the sea. Hyperion, New York

Ballard RD (1998) High-tech search for Roman shipwrecks. Natl Geogr 193(4):32–41

Ballard RD, Coleman DF, Rosenberr GD (2000) Further evidence of abrupt drowning of the Black Sea shelf. Mar Geol 170(3–4):253–261

Barbin V, Ramseyer K, Burns SJ, Decrouez D, Maier JL, Chamay J (1991) Cathodluminescence signature of white marble artefacts. Mater Res Soc Symp Proc 185:299–308

Barbin V, Ramseyer K, Decrovez D, Burns SJ, Chambay J, Maier JL (1992) Cathodluminescence of white marbles: an overview. Archaeometry 34(2):175–183

Barker RD (1992) A simple algorithm for electrical imaging of the subsurface. First Break 10(2):53–62

Barnes IL, Gramlich JW, Diaz MG, Brill RH (1978) The possible change of lead isotope ratios in the manufacture of pigments: a fractionation experiment. In: Carter GF (ed) Archaeological chemistry, II. American Chemical Society, Washington, DC, pp 273–279

BaVetto, Villeneuve VG, Schvoerer M, Bechtel F, Herz N (1999) Investigation of electron paramagnetic resonance peaks in some powdered white marble. Archaeometry 41(2):253–265

Baxter MJ, Jackson CM (2001) Variable selection in artefact compositional studies. Archaeometry 43:253–268

Bearrs EC (1966) Hard luck ironclad. Louisiana State University Press, Baton Rouge

Bettis EA III (1995) Archaeological geology of the archaic period in North America. Geological Society of America Special Paper 297. Geological Society of America, Boulder, CO

Birkeland PW (1984) Soils and geomorphology. Oxford University Press, New York

Bishop RL (1992) Comments on section II: variation, characterization of ceramic pastes in archaeology. In: Neff H (ed) Monographs in world archaeology, no 7. Prehistory Press, Madison, pp 167–170

Blackwell B, Schwarz H (1993) Archaeochronology and scale. In: Stein JK, Linse AR (eds) Effects of scale on archaeological and geoscientific perspectives. Geological Society of America, Special Paper 283, Geological Society of America, Boulder, CO

Blair TC, McPherson JG (1999) Grain-size and textural classification of coarse sedimentary particles. J Sediment Res 69:6–19

Blanchar RW, Rehm G, Caldwell AC (1965) Sulfur in plant remains by digestion with nitric and perchloric acid. Proc Soil Sci Soc Am 29:71–72

Blatt H (1982) Sedimentary Petrology. WH Freeman, San Francisco

Bond G, Broecker W, Johnson S, McMamus J, Labeyrie L, Jouzel L, Bonani G (1993) Correlation between climate records from North Atlantic sediments and Greenland ice. Nature 365: 143–147

Bordes F (1972) A Tale of Two Caves. Harper and Row, New York

Boris L de, Melentis J (1991) Age et position phylétique du Crâne de Petralona (Grèce). Les Premiers Européens. Editions du CTHS, Paris

Bosinski G (1992) Eiszeitjäger im Neuwieder Becken. Archäologie an Mittelrhein und Mosel. Band 1. Gesellschaft für Archäologie am Mittelrhein und Mosel EV. Archäologische Denkmalpflege Amt, Koblenz [incomplete]

Bozarth SR (1986) Morphologically distinctive Phaseolus, Cucurbita, and *Helianthus annuus* phytoliths. Plant opal phytolith analysis in archaeology and paleoecology. In: Rovner I (ed) Occasional papers of the Phytolitharian, I. North Carolina State University, Raleigh, pp 56–66

Bozarth SR (1993) Maize (*Zea Mays*) cob phytoliths from a central Kansas Great Bend Aspect archaeological site. Plains Anthropol 38:279–286

Brady N (1990) The nature and properties of soils. MacMillan, New York

Breiner S (1973) Applications manual for portable magnetometers. Geometrics, Palo Alto, CA, 58 pp

Briard J (1979) The bronze age in barbarian Europe: from the megaliths to the Celts. Translated by Mary Turton. Routledge and K Paul, London

Brook GA, Coward JB, Marais E (1996) Wet and dry periods in the southern African summer rainfall zone during the last 300 kyr from speleothem, tufa and sand dune age data. In: Heine K (ed) Paleoecology of Africa and the surrounding islands, vol 24. Balkema, Lisse, Netherlands, pp 147–158

Brown DA (1984) Prospects and limits of a phytolith key for grasses in the central United States. J Archaeol Sci 11 (4):345–368

Bruno M, Lazzarini L (1995) Discovery of the Sienese provenance of brecci dorata and breccia giallo fibrosa, and the origin of breccia rossa appennica. Archeomateriaux: Marbres et autres roches. Acts de la Ive Conférence Internationale, ASMOSIA IV CRPAA Pub Presse Universitaires Bordeaux, Bordeaux

Bryant VM Jr (1974) The role of coprolyte analysis in archaeology. Bull Tex Archaeol Soc 45:1–28

Bryant VM Jr (1978) Palynology: a useful method for determining paleoenvironmental. Tex J Sci 30:25–42

Bryant VM Jr (1989) Pollen: natures fingerprints of plants. In: 1990 Yearbook of science and the future. Encyclopedia Britannica Inc, Chicago

Bryant VM Jr, Holloway RG (1983) The role of palynology in archaeology. In: Schiffer MB (ed) Advances in archaeological method and theory. Academic Press, New York, pp 191–223

Bryson RA (1988) What the climatic past tells us about the environmental future. Earth '88, changing geographic perspectives. National Geographic Society, Washington

Buckley DE, Mackinnon WG, Cranston RE, Christian HA (1994) Problems with piston core sampling: mechanical and geochemical diagnosis Mar Geol 117(1–4); 95–106

Bukry D (1979) Comments on opal phytoliths and stratigraphy of Neogene silica-flagellates and coccoliths at Deep Sea Drilling Project Site 397 off northwest Africa. In: Shamback JD (ed) Initial reports of the Deep Sea Drilling Project 49. US Government Printing Office, Washington, DC, pp 977–1009

Burger HR (1992) Exploration geophysics of the shallow subsurface. Prentice Hall, Englewood Cliffs

Busch RM (ed) (1996) Laboratory manual in physical geology, 4th edn. Prentice Hall, Upper Saddle River, NJ

Butzer KW (1982) Archaeology as human ecology. Cambridge University Press, Cambridge

Caldwell AC (1965) Sulfur in plant materials by digestion with nitric and perchloric acid. Soil Sci Soc Am Proc 29:71–72

Cannon MD (2001) Archaeological relative abundance, sample size and statistical methods. J Archaeol Sci 28:185–195

Capitani A de (1993) Maur ZH – Schifflände. Die Tauchuntersuchungen der Ufersiedlung 1989 bis 1991. Jahrb Schweiz Ges Ur- Frühgesch 76:45–70

Carey C (1999) Secrets of the sacrificed. Discov Archaeol 1(4):46–53

Carr TL, Turner MD (1996) Investigating regional lithic procurement using multi-spectral imagery and geophysical prospection. Archaeol Prospection 3:109–127

Carver RE (ed) (1971) Procedures in sedimentary geology. Wiley Interscience, New York

Caton-Thompson G, Gardner EW (1934) The Desert Fayum. The Royal Anthropological Institute of Great Britain and Ireland, London

Catt JA, Weir AH (1976) The study of archaeologically important sediments by petrographic techniques. In: Davidson DA, Shackley MC (eds) Geoarchaeology. Westview Press, Boulder, CO

Cavanagh WG, Hirst S, Litton CD (1988) Soil phosphate, site boundaries and change point analysis. J Field Archaeol 15:67–83

Champion T, Gamble C, Shennan S, Whittle A (1984) Prehistoric Europe. Academic Press, London

Chayes F (1954) The theory of thin-section analysis. J Geol 62:92–101

Chayes F (1956) Petrographic modal analysis. Wiley, New York

Cherry J (1988) Island origins. In: Barry Cunliffe (ed) Origins. The Dorsey Press, Chicago

Childe VG (1951) Man makes himself. New American Library, New York

Clark AJ (1986) Archaeological geophysics in Britain. Geophysics 51:1404–1413

Clark AJ (1990) Seeing beneath the soil. BT Batsford Ltd, London

Clarke GR, Beckett P (1971) The study of soil in the field. Oxford University Press, Oxford

Clottes J (2001) La Grotte Chauvet – L'Art des Origines. Editions de Sevil, Paris

Clottes J, Courtin J (1996) The cave beneath the sea: paleolithic images at Cosquer. HN Abrams, New York

Collinson JD (1996) Alluvial sediment. In: Reading HG (ed) Sedimentary environments: processes, facies and stratigraphy, 3rd edn. Blackwell Science, Oxford, pp 37–82

Comer DC (1998) Discovering archaeological sites from space. CRM (21)5:9–11

Conte DJ, Thompson DJ, Moses LL (1997) Earth science: an integrated perspective, 2nd edn. Wm C Brown, Dubuque

Corti ECC (1951) The destruction of Pompeii and Herculaneum. Routledge and Paul, London

Courty MA (1992) Soil micromorphology in archaeology. In: Pollard EM (ed) Proceedings of the British Academy 77:39–59. Oxford University Press, London

Courty MA, Goldberg PA, Macphail RI (1989) Soil and micromorphology in archaeology. Cambridge University Press, Cambridge

Cowgill GL (1964) The selection of samples from large sherd collections. Am Antiquity 29(4): 467–473

Craddock P, Gurney D, Pryor F, Hughes M (1985) The application of phosphate analysis to the location and interpretation of archaeological sites. Archaeol J 142:361–376

Craddock PT, Cowell MR, Leese MN, Houghes MJ (1983) The trace element composition of polished flint axes as an indicator of source. Archaeometry 25:135–163

Cremeens DL, Hart JP (1995) On chronostratigraphy, pedostratigraphy, and archaeological context. Pedological perspectives in archaeological research. SSSA Special Publication 44. Soil Science Society of America, Madison, WI

Cunliffe B (1988) Aegean civilization and barbarian Europe. In: Barry Cunliffe (ed) Origins. The Dorsey Press, Chicago

Dalan RA, Musser JM Jr, Stein JK (1992) Geophysical exploration of the Shell Midden. In: Stein JK (ed) Deciphering a shell midden. Academic Press, San Diego

Daniel G (1967) The origins and growth of archaeology. Penguin Books, Baltimore

Davidson DA, Shackley ML (eds) (1976) Geoarchaeology. Westview Press, Boulder, CO

Davis JC (1986) Statistics and data analysis in geology. Wiley, New York

Davis M (1969) Palynology and environmental history during the Quaternary period. Am Sci 57(3):317–322

Davis WM (1899) The geographical cycle. Geogr J 14:481–504

Day PR (1965) Particle fractionation and particle size analysis. In: Black CA, Evans DD, Dinauer RC (eds) Methods of soils analysis, part I agronomy. American Society of Agronomy, Madison, WI, pp 545–567

De Bruin M, Korthoven TJM, van der Steen AJ, Huitman JPW, Duin RPW (1976) The use of trace element concentrations in the identification of objects. Archaeometry 18:75–83

Del Monte M, Ausset P, Lefevre RA (1998) Traces of ancient colors on Trajan's Column. Archaeometry 40(2):403–412

De Lumley H (1969) A paleolithic camp at Nice. Sci Am 220(5):42–50

Descantes C, Neff H, Glasscock MD, Dickinson WR (2001) Chemical characterization of Micronesian ceramics through instrumental neutron activation analysis. A preliminary provenance study. J Archaeol Sci 28:1185–1190

Dort W Jr (1978) Geologic basis for assignment of >18,000 year age to cultural material at the Shriver site, Daviss County, northwestern Missouri. Abstracts. Geological Society of America Annual Meeting, Geological Society of America, Boulder, CO

Drennan RD (1996) Statistics for archaeologists. Plenum, New York

Dunnell RC, McCutcheon PT, Ikeya M, Toyoda S (1994) Heat treatment of Mill Creek and Dover cherts on the Malden Plain, southeast Missouri. J Archaeol Sci 21(1):79–89

Earth Observing System (EOS) (1987) HIRIS, High Resolution Imaging Spectrometer: science opportunities in the 1990s, vol IIC. National Aeronautical and Space Administration, Washington

Easterbrook DS (1998) Surface processes and landforms. Macmillan, New York

Edgerton H (1986) Sonar images. Prentice Hall, Englewood Cliffs

Ehrenberg CG (1854) Mikrogeologie. Leopold Voss, Leipzig

Eidt RC (1973) A rapid chemical test for archaeological site surveying. Am Antiquity 38:206–210

Eidt RC (1977) Detection and examination of anthrosols by phosphate analysis. Science 197:1327–1333

Eidt RC (1985) Theoretical and practical considerations in the analysis of anthrosols. In: Rapp G Jr, Gifford JA (eds) Archaeological geology. Yale University Press, New Haven, pp 155–190

Eidt RC, Woods WI (1974) Abandoned settlement analysis: theory and practice. Field Test Associates, Shorewood, WI

Eighmy JL, Sternberg RS (eds) (1990) Archaeomagnetic dating. University of Arizona Press, Tucson

El-Gammili MM, El-Mahmondi AS, Osman SS, Hassareem AG, Metwaly MA (1999) Geoelectric resistance scanning on parts of Abydos cemetery region, Sohag Govenorate, upper Egypt. Archaeol Prospection 6:225–239

Ellwood BB, Harrold FB, Marks AE (1994) Site identification and correlation using geoarchaeological methods at the Cabeho do Porto Marinlo (CPM) locality Rio Maior, Portugal. J Archaeol Sci 21(6):779–784

Ellwood BB, Peter DE, Balsam W, Schieber J (1995) Magnetic and geochemical variations as indicators of paleoclimate and archaeological site evolution: examples from 41 TR68, Fort Worth, Texas. J Archaeol Sci 22(3):409–415

Emery WB (1961) Archaic Egypt. Pelican Books, Baltimore

Faegri K, Iverson J (1975) Textbook of pollen analysis, 3rd edn. Hafner Publishing, New York

Fagan BM (1994) Quest for the past, 2nd edn. Waveland Press, Prospect Heights, IL

Fagan BM (1999) Archaeology: theories, methods and practice. Thames and Hudson, London

Farrand WR (1975) The analysis of quaternary cave sediments. World Archaeol 10:290–301

Faure G (1986) Principles of isotope geology. Wiley, New York

Feder KL (1997) Indians and archaeologists. Skeptic 5(3):44–51

Ferring CR, Perttula TK (1987) Defining the provenience of red-slipped pottery from Texas and Oklahoma by petrographic methods. J Archaeol Sci 14:437–456

Fiedel SJ, Southon JR, Brown TA (1995) The GISP ice core record of volcanism since 7000 B.C. Science 267:256–258

Flenniken JJ, Garrison EG (1975) Thermally altered novaculite and stone tool manufacturing techniques. J Field Archaeol 2:125–131

Folk RL (1968) Petrology of sedimentary rocks. Hemphill, Austin, TX

Follmer LR (1985) Surficial geology and soils of the Rhoads archaeological site near Lincoln, Illinois. Am Archaeol 5(2):150–160

Ford WE (1918) The growth of mineralogy from 1818–1918. Am J Sci 46:240–254

Fowkles EB, Gnanadesikan R, Kettenring JR (1988) Variable selection in clustering. J Classif 5:205–288

Freestone IC, Middleton AP (1987) Mineralogical applications of the analytical SEM in archaeology. Mineral Mag 51:21–31

Függel E (1982) Microfacies analysis of limestones. Springer, Berlin Heidelberg New York.

Gagliano SM, Pearson CE, Weinstein RA, Wiseman DE, McClendon CE (1982) Sedimentary studies of prehistoric archaeological sites. Coastal Environments Inc, Baton Rouge, LA

Galehouse JS (1971) Point counting. In: Carver R (ed) Procedures in sedimentary petrology. Wiley Interscience, New York

Gall J-C (1983) Ancient sedimentary environments and the habitats of the living organisms: introduction to paleoecology. Springer, Berlin Heidelberg New York

Garrels RM, Mackensie FT (1971) Evolution of sedimentary rocks. WW Norton, New York

Garret EM (1986) A petrographic analysis of Black Mesa ceramics. In: Plog S (ed) Spatial organization and trade. Southern Illinois Press, Carbondale, pp 114–142

Garrison EG (1998) Radar prospection and cryoprobes – early results from Georgia. Archaeol
Prospection 5:57–65

Garrison EG (1999) A history of engineering and technology, 2nd edn. CRC Press, Boca Raton,
FL

Garrison EG (2000) A geoarchaeological study of Prisoners Harbor, Santa Cruz Island, Northern Channel Islands, California – past sea levels and deltaic deposits of an insular stream.
Proceedings of the 5th California Islands Symposium, 29 March–1 April 1999, Department
of the Interior, Minerals Management Service, OCS Study, MMS 99–0038

Garrison EG (2001) Physics and archaeology. Phys Today 54:32–36

Garrison EG, Baker JG, Thomas DH (1985) Magnetic prospection and discovery of Mission Santa
Catalina de Guale. J Field Archaeol 12(3):299–313

Gè T, Courty M-A, Matthews W, Wattez J (1993) Sedimentation formation processes of occupation surfaces. Formation Processes in archaeological contexts. In: Goldberg P, Nash PT,
Petraglia MD (eds) Monographs in world archaeology, no 17. Prehistory Press, Madison, WI,
pp 143–63

Gee GW, Bauder JW (1986) Particle size analysis. Methods of soil analysis, part I: physical and
mineralogical methods, 2nd edn. Klute A (ed) Agronomy No 1, American Society of
Agronomy, Madison, WI, pp 383–411

Gerrard MC (1991) Sedimentary petrology and the archaeologist: the study of ancient
ceramics. In: Morton AC, Todd SP, Haughton DW (eds) Developments in sedimentary provenance studies. Geological Society Special Publication 57. Geological Society of America,
Boulder, CO, pp 189–197

Gibbons A (1991) A new look for archaeology. Science 252:918–920

Gladfelter BG (1977) Geoarchaeology: the geomorphologist and archaeology. Am Antiquity
42:512–538

Glasscock MD (1992) Characterization of archaeological ceramics by neutron activation analysis and multi-variate statistics. In: Neff H (ed) Chemical characterization of ceramic pastes
in archaeology. Monographs in world archaeology, no 7. Prehistory Press, Madison, WI,
pp 11–26

Glasscock MD, Neff H, Stryker KS, Johnson TN (1994) Sourcing archaeological obsidian by
abbreviated NAA procedure. J Radioanal Nucl Chem Articles 180:29–35

Godfrey-Smith DI, Huntley DJ, Chen WH (1988) Optical dating of quartz and feldspar sediment
extracts. Q Sci Rev 7:373–380

Goffer Z (1980) Archaeological chemistry: a sourcebook on the applications of chemistry to
archaeology. John Wiley and Sons, New York

Goldberg PA (1995) Microstratigraphy, micromorphology site formation processes, soils. In: The
practical impact of science on field archaeology: maintaining long-term analytical options.
A workshop on Cyprus, 22 and 23, July 1995. The Weiner Laboratory of the American School
of Classical Studies, Athens

Goldberg P (1999) Late-terminal classic Maya pottery in northern Belize: a petrographic analysis of sherd samples from Colha and Kichpanha, Iceland, HB. J Archaeol Sci 26(8):951–
966

Goodman D, Conyers LA (1997) Ground penetrating radar: an introduction for archaeologists.
Altamira Publishing, Walnut Creek, CA

Gorsuch R (1984) Factor analysis. L Erlbaum Associates, Hillsdale, NJ

Gramlich JW, Barnes IL, Diaz MG, Brill RH (1978) The possible change of lead isotope ratios in
the manufacture of pigments. In: Carter GF (ed) Archaeological chemistry, II. American
Chemical Society, Washington, DC, pp 273–279

Gratuze B, Goivagnoli A, Banadon JN, Telouk P, Imbert JL (1993) Apport de la méthode ICP-MS
couplée à l'ablation laser pour la caractérisation des archéomatériaux. Rev Archeometrie
17:89–104

Grebothé D, Lassau G, Ruckstuhl M, Seifert (1990) Thayngen SH-Weier: Trockeneissondierung 1989. Jahrb Schweiz Ges Ur- Frühgesch 73:167–175

Green AJ (1981) Particle-size analysis. In: McKeague JA (ed) Manual on soil sampling and methods of analysis. Canadian Society of Soil Science, Ottawa, pp 4–29

Greenough JD, Dobosi G, Owen JV (1999) Fingerprinting ancient Egyptian quarries: preliminary results using laser ablation microprobe-inductively coupled plasma-mass spectroscopy. Archaeometry 41(2):227–238

Griffiths GH, Barker RD (1994) Electrical imaging in archaeology. J Archaeol Sci 21:153–158

Grob A (1896) Beitrage zur Anatomie der Epidermis der Gramineenblatter. Bibl Bot 36:1–63

GSA Today (2002) 2001 medals and awards. Geol Soc Am 12:21–22

Guineau B, Lorblanchet M, Gratuze B, Dulin L, Roger P, Akrich R, Muller F (2001) Manganese black pigments in prehistoric paintings: the case of the black frieze of Pech Merle (France). Archaeometry 43:211–225

Gvirtzman G, Weider M (2001) Climate of the last 53,000 years in the eastern Mediterranean, based on soil-sequence stratigraphy in the coastal plain of Isrl. Q Sci Rev 20:1827–1849

Hach water analysis handbook, 2nd edn (1992) Hach Co, Loveland, CO

Hadorn P (1994) Saint-Blaise/Bains des Dames. 1. Palynologie d'un site néolithique et historique de la végétation der derniers 16,000 ans. Musée cantonal d'archéologie, Neuchâtel (Archéologie Neuchâteloise 18)

Hampton RE (1994) Introductory Biological Statistics. Wm C Brown Publishers, Dubuque, IA

Harbottle G (1976) Activation analysis in archaeology. Radiochemistry 3:33–72

Hassan FA (1988) Fluvial systems and geoarchaeology in arid lands with examples from north Africa, the Near East and the American southwest. In: Stein JK, Farrand WR (eds) Archaeological sediments in context. Center for the Study of Man, Orono, ME

Hatcher RD (1978) Tectonics of the western Piedmont and Blue Ridge, southern Appalachians. Am J Sci 278:276–304

Hawkes J (1995) Discovering statistics. Quant Publishing, Charleston

Hay RL (1976) Geology of Olduvai Gorge: a study of sedimentation in a semiarid Basin. University of California Press, Berkeley, CA

Heaney PJ (1995) Moganite as an indicator for vanished evaporites: a testament reborn? J Sediment Res A65:633–638

Heard P (1996) Cathodluminescence-interesting phenomenon on useful technique? Microsc Anal 16:25–27

Henderson J (2000) The science and archaeology of materials. Routledge, London

Hendy CH (1971) The isotopic geochemistry of speleothems, 1. The calculation of the effects of different modes of formation on the isotopic composition of speleothems and their applicability as paleoclimatic indicators. Geochim Cosmochim Acta 35:801–824

Herbich T (1993) The variations of shaft fills as the basis of the estimation of flint mine extent: a Wierzvica study. Archaeol Polona 31:71–82

Herbich T (1995) The application of geophysical methods to the study of prehistoric flint mines. In: Science and the site: evaluation and conservation. Beavis J, Barker K (eds) Bournemouth University School of Conservation Sciences Occasional Paper 1. Bournemouth, UK

Hermes OD, Luedtke BE, Ritchie D (2001) Melrose green rhyolite: Its geologic setting and petrographic and geo-chemical characteristics. J Archaeol Sci 28:913–928

Herz N (1987) Carbon and oxygen isotopic ratios: a data base for classical Greek and Roman marble. Archaeometry 29(1):35–43

Herz N, Garrison EG (1998) Geological methods for archaeology. Oxford University Press, Oxford

Herz N, Holbrow KA, Sturman SG (1999) Marble sculpture in the National Gallery of Art. Archéomatériaux, marbres et autre roches. ASMOSIA IV Schroerer M (ed) Presses de Universitaires de Bordeaux. p 110

Hesse A (1999) Multi-parametric survey for archaeology. How and why, or how and why not? J Appl Geophys 41 (2–3):157–168

Higgins MD, Higgins R (1996) A geological companion to Greece and the Aegean. Cornell University Press, Ithaca

Hoard RJ, Holen SR, Glasscock MD, Neff H, Elam JM (1992) Neutron activation analysis of stone from the Chadron formation and a Clovis site on The Great Plains. J Archaeol Sci 19:655–665

Hochuli S (1994) Unter den "Bahn 2000"; Gefrierkern Bohrung in Kanton Zug. Archaöl Schweiz 17(1):25–30

Hodgson JM (1978) Soil sampling and soil description. Oxford University Press, London

Holcomb DW (1996) Shuttle imaging radar and archaeological survey in the China Taklamakan desert. J Field Archaeol 19(1):129–138

Holley GR, Dalan RA, Smith PA (1993) Investigations in the Cahokia Site Grand Plaza. Am Antiquity 58(2):306–319

Howell JM (1987) Early farming in northwestern Europe. Sci Am 237:118–126

Hsü K (1995) The geology of Switzerland. Princeton University Press, Princeton

Inbar M, Hubp JL, Villers Riuz L (1994) The geomorphological evolution of the Paricutin cone and lava flow, Mexico, 1943–1990. Geomorphology 9:57–76

Indorante SJ, Follmer LR, Hammer RD, Koenig PG (1990) Particle-size analysis by a modified pipette procedure. Soil Sci Soc Am J 54:560–563

Ikeya M (1985) Electron spin resonance. In: Rufter NW (ed) Dating methods of Pleistocene deposits and their problems. Geoscience Canada reprint series 2. Geological Association of Canada, Toronto

Ivester AH, Leigh DS, Godfrey-Smith D (2001) Chronology of source-bordering dunes in the coastal plain of Georgia, USA. Quat Res 55(3):293–302

Jarvis KE, Gray AL, Houk RS (1992) Handbook of inductively coupled plama mass spectrometry. Blackie, Glasgow

Jelinek A, Farrand WR, Haas G, Horowitz A, Goldberg AP (1973) New excavations at Tabun Cave, Mount Carmel, Israel: preliminary report. Paléorient I:151–183

Jercher M, Pring A, Jones PG, Raven MD (1998) Rietveld X-ray diffraction and X-ray fluorescence analysis of Australian aboriginal ochres. Archaemetry 40:383–401

Johanssen A (1938) A descriptive petrography of the igneous rocks. University of Chicago Press, Chicago

Johnson NL, Leone FC (1964) Statistics and experimental design in engineering and the physical sciences, vol I. Wiley, New York

Jones AA (1991) X-ray fluorescence analysis. In: Smith KA (ed) Soil analysis: modern instrumental techniques, 2nd edn. Marcel Dekker, New York, pp 287–324

Jones RL (1964) Note on occurrence of opal phytoliths in some Cenozoic sedimentary rocks. J Paleontol 38:773–775

Jurmain R, Nelson H, Kilgore L, Trevathan W (2000) Introduction to physical anthropology. Wadsworth, Belmont, CAL

Kallithrakas-Kontos N, Katsamos AA, Aravantinos A (1993) Study of ancient Greek copper coins from Nikopolis (Epirus) and Thessaloniki (Macedonia). Archaeometry 35:265–278

Kamei H, Nishimura Y, Komatsu M, Saito M (1992) A new instrument: a three-component Fluxgate gradiometer. Abstracts of the International Archaeometry Symposium, Los Angeles, CA, p 71

Kamilli D, Sternberg A (1985) New approaches to mineral analysis in ancient ceramics. In: Rapp G Jr, Gifford JA (eds) Archaeological geology. Yale University Press, New Haven, pp 313–330

Kamilli D, Lamberg-Karlovsky CC (1979) Petrographic and electron microprobe analysis of ceramics from Tepe Yalya. Archaeometry 21:47–60

Kearey P, Brooks M (1991) An introduction to geophysical exploration. Blackwell Science, Oxford

Keller EA, Pinter N (2002) Active tectonics: earthquakes, uplift and landscape, 2nd edn. Prentice Hall, Upper Saddle River, NJ

Kempe DRC, Harvey AP (eds) (1983) The petrology of archaeological artifacts. Clarendon Press, Oxford

Kempthorne O, Allamaras RR (1986) Errors and variability of observations. Methods of soil analysis, part i agronomy. Soil Science Society of America, Madison, WI, pp 1–31

Kennett DJ, Neff H, Glasscock MD, Mason AZ (2001) Interface – archaeology and technology. A geochemical revolution: inductively coupled plasma mass spectroscopy. SAA Archeol Record 1:22–26

Kennett JP, Baldauf JG, Behl R (1994) Proc ODP, International Reports, 146 (Part 2), Ocean Drilling Program, College Station, Texas

Khakimov Akh R (1957) Artificial freezing of soils. Theory and practice. Academy of Sciences of the USSR Permafrost Institute in VA Obruche. Translation by Israel Program in Scientific Translations (1966)

Kish L (1965) Survey sampling. John Wiley and Sons, New York

Klecka WR (1975) Discriminate analysis. In: Nie NH, Bent DH, Hull CH (eds) Statistical package for the social sciences (SPSS). McGraw-Hill, Englewood Cliffs, pp 434–467

Klockenkämper R, Bubert H, Hasler K (1999) Detection of near-surface silver enrichment on Roman coins. Archaeometry 41(2):311–320

Kraft JC (1994) Archaeological geology. Geotimes 39:12–13

Krings M, Stone A, Schmitz RW, Krainitzi H, Stoneking M, Pääbo S (1997) Neanderthal DNA sequences and the origin of modern humans. Cell 90:19–30

Kruk J (1980) The Neolithic settlement of southern Poland. British Archaeological Reports S93, Oxford

Krumbein WC, Pettijohn FJ (1938) Manual of sedimentary petrography. Appleton and Century, New York

Kubiena ML (1953) The soils of Europe. Murby, London

Kuhn RD, Sempowski ML (2001) A new approach to dating the League of the Iroquois. Am Antiquity 66:301–314

Kukla G, Heller F, Liu X, Xu M, Liu TS, An ZS (1988) Pleistocene climates in China dated by magnetic susceptibility. Geology 16:811–814

Kuper R, Lohr H, Lüning J, Stehli P, Zimmerman A (1977) Der Band Keramische Siedlungsplatz Langweiler, 9, Gem. Aldenhoven, Kr. Düren. Rheinische Ausgrabungen 18. Rheinland Verlag, Bonn

Kzanowski WJ (1988) Principles of multivariate analysis. Clarendon Press, Oxford

Lamberg-Karlovsky CC (1974) Excavations at Tepe Yalya. In: Willey GR (ed) Archaeological researches in retrospect. Winthrop, Cambridge, pp 269–292

Lassau G, Riethmann P (1988) Trockeneissondierung, ein Prospektionsverfahren im Seeuferbereich. Jahrb Schweiz Ges Ur- Frühgesch 71:241–247

Lazerwitz B (1968) Sampling theory and procedures in methodology in social research. McGraw-Hill, New York

Leach F, Manley B (1982) Minimum Mahalanobis distance functions and lithic source characterization by multi-element analysis. N Z J Archaeol 4:77–109

LeBlanc G (2001) A review of EPA sample preparation techniques from organic compounds of liquid and solid samples. LC GC 19:1120–1121

LeBorgne P (1960) Influence du fer sur les propriétés magnétiques du sol et sur celles du schiste et du granite. Ann Geophys 6(2):159–195

Leigh DS (1998) Evaluating artifact burial by eolian versus bioturbation processes, South Carolina sandhills, USA. Geoarchaeo 13:309–330

Les Dossiers d'Archéologie (1991) 162:65–81

Leudtke BE (1992) An archaeologist's guide to chert and flint. University of California Press, Los Angeles, CA

Leute U (1987) Archaeometry. VCH Verlag, Weinheim

Levin HL (1988) The earth through time, 3rd edn. Saunders College Publishing, Philadelphia

Levin SA (1992) The problem of pattern and scale in ecology. Ecology 73:1943–1967

Linderholm J, Lundberg E (1994) Chemical characterization of various archaeological soil samples using main and trace elements determined by inductively coupled plasma atomic emission spectrometry. J Archaeol Sci 21:303–314

Littman SB (2000) Pleistocene/Holocene sea level change and lithostratigraphy of the Georgia Bight: a geoarchaeological study. Masters Thesis, Department of Geology. University of Georgia, Athens

Lombard JP (1987) Provenance of sand temper in Hohokam ceramics. Geoarchaeology 2:91–119

Lotter AF, Renberg I, Hansson H, Stockli R, Sturm M (1997) A remote controlled freeze corer for sampling unconsolidated surface sediments. Aquat Sci 59(4):295–303

Loy TH, Dixon EJ (1998) Blood residues on fluted points from eastern Beringia. Am Antiquity 63:21–46

Lucas A, Harris JR (1962) Ancient Egyptian materials and industries. Edward Arnold Publishers Ltd, London

Lyell C (1863) The geological evidence for the antiquity of man. J Murray, London

Mackensie WS, Donaldson CH, Guilford C (1982) Atlas of igneous rocks and their textures. Addison Wesley Longman, Harlow

Maggetti M (1982) Phase analysis and its significance for technology and origin. In: Olin JS, Franklin AD (eds) Archaeological ceramics. Smithsonian Institution, Washington, DC, pp 121–133

Maher LJ (1981) Statistics for microfossil concentration measurements employing samples spiked with marker grains. Rev Paleobot Palynol 32:153–191

Malainey ME, Przybylski R, Sheriff BL (1999) Identifying the former contacts of late precontact period pottery vessels from western Canada using gas chromatography. J Archaeol Sci 26(4):425–438

Mallory-Greenough LM, Greenough JD (1998) New data for old pots: trace-element characterization of ancient Egyptian pottery using ICP-MS. J Archaeol Sci 25(1):85–97

Mannoni L, Mannoni T (1984) Marble. Facts on File Publications, New York

Marmet E, Bina M, Fedoroffand N, Tabbagh A (1999) Relationships between human activity and the magnetic properties of soils: a case study in the medieval site of Roissy-en-France. Archaeol Prospection 6:161–170

Marshall A (1999) Magnetic prospection at high resolution: survey of large silo-pits in Iron Age enclosures. Archaeol Prospection 6:11–29

Martin PD (1971) The last 10,000 years. University of Arizona Press, Tucson

Maslin MA (2000) North Atlantic iceberg armadas. Sci Spectra 22:40–50

Matson FR (1960) The quantitative study of ceramic materials. In: Heizer RF, Cook SF (eds) The application of quantitative methods in archaeology. Viking fund publications in anthropology no 28. Wenner Gren Foundation, New York, pp 43–51

Matson FR (1961) Ceramics and man. Viking fund publications in anthropology no 41. Methuen, London

Mazeran R (1995) Les brèches exploitées comme marbre dans le Sud-Est de la France a l'époque romaine. Archéomatériaux: Marbres et autres roches. Actes de la Iv^e Conférence internationale, ASMOSIA IV. CRPAA Pub Presses Universitaires de Bordeaux, Bordeaux

McCauley JF, Schaber GG, Breed CS, Grolier MJ, Haynes CU, Issawi B, Elachi C, Blom R (1982) Subsurface valleys and geoarchaeology of the Eastern Sahara revealed by Shuttle Radar. Science 218:1004

McKay ED (1979) Stratigraphy of Wisconsin and older loesses in southwestern Illinois. Ill Geol Survey Guidebook No 14:37–67

McMorrow J (1995) Multispectral remote sensing of archaeological sites: NERC airborne thematic mapper images of the Oxford flood plain. In: Beavis J, Baker K (eds) Science and the site. Bournemouth University Occasional Papers 1, Bournemouth

Meats C (1996) An appraisal of the problems involved in three-dimensional ground penetrating radar imaging of archaeological features. Archaeometry. 38(2):359–379

Melas EM (1985) The islands of Karpathos, Saros and Kasos in the Neolithic and bronze age. Stud Mediterranean Archaeol LXVII:15–24

Meriweather DA (1999) Freezer anthropology: new uses for old blood. Philos Trans R Soc Lond B 354:121–129

Meschel SV (1980) Chemistry and archaeology: a synergism. Chemtech 10(7):404–410

Mikhail EH, Briner GP (1978) Routine particle size analysis using sodium hypochlorite and ultra-sonic dispersion. Aust J Soc Res 14:241–244

Milson J (1996) Field geophysics, 2nd edn. Wiley, London

Miskimmin BM, Curtis PJ, Schindler DW, Lafaut N (1996) A new hammer-driven freeze corer. J Paleolimnol 15(3):265–269

Monttana A, Grespi R, Liborio G (1978) Simon and Schuster's guide to rocks and minerals. Simon and Schuster Inc, New York

Morell V (1995) Who owns the past. Science 268:1424–1426

Mulaik S (1972) The foundations of factor analysis. McGraw-Hill, New York

Mulholland SC, RappG Jr (eds) (1992) Phytolith systematics: emerging issues. Plenum Press, New York, pp 1–13

Mullins CE (1974) The magnetic properties of the soil and their application to archaeological prospecting. Archaeo Phys 5:144–148

Murphy J, Riley JP (1962) A modified single solution for the determination of phosphorus in natural waters. Anal Chim Acta 27:21–26

Neff H, Glasscock MD (1995) The state nuclear archaeology in North America. J Radioanal Nucl Chem 196:275–286

Neubauer W (1999) Geophysikalische Prospektion in der Archäologie. Institut für Ur- und Frühgeschichte, Wien

Nikischer T (1999) Modern mineral identification techniques, part I WDS and EDS. Mineral Rec 30:297–300

Noel M, Xu B (1991) Archaeological investigation by electrical resistive tomography: a preliminary study. Geophys J Int 107:95–102

Nur A, Ron H (1997) Earthquake-inspiration for Armageddon. Bibl Archaeol Rev 23(4): 49–55

Odum EP (1995) Ecology: science and society. Sinauer, Sunderland, MA

Olson GW (1981) Soils and the environment. Chapman and Hall, New York

Orliac M (1975) Empreintes au latex des coupes du gisement magdalénien de Pincevent: technique et premier résultats. Bull Soc Prehist Fr 72:274–276

Osterwalder C, André R (1980) La Suisse Préhistorique, vol 1. Editions 24 Heures, Lausanne.

Oviatt CG, Swinehart JB, Wilson JR (1997) Dryland landforms, hazards and risks. In: Busche RM (ed) Laboratory manual for physical geology, 4th edn. Prentice Hall, Upper Saddle River, NJ

Owens DL (1997) A feasibility study for phytolith research in the southeast from Skull Shoals in the Oconee National Forest and Skidaway Island, Georgia. Masters thesis. Department of Geology. University of Georgia, Athens

Palmer AN (1991) Original morphology of limestone caves. GSA Bull 103:1–21

Palmer JW, Hollander MG, Rodgers PSZ, Benjamin TM, Duffy CJ, Lambert JB, Brown JA (1998) Pre-Columbian metallurgy: technology, manufacture, and microprobe analyses of copper bells from the greater Southwest. Archaeometry 40(2):361–382

Pansu M, Gautheyrou J, Loyer J-Y (2001) Soil analysis: sampling, instrumentation and quality control. Translated by VAK Sarma. Balkema, Lisse, Netherlands

Papageorgiou I, Baxter MJ, Cau MA (2001) Model-based cluster analysis of artefact compositional data. Archaeometry 43:571–588

Parker SP (ed) (1994) McGraw-Hill dictionary of geology and mineralogy. McGraw-Hill, New York

Patel SB, Hedges REM, Kilner JA (1998) Surface analysis of archaeological obsidian by SIMS. J Archaeol Sci 25(10):1047–1054

Patella D, Hesse A (eds) (1999) Special issue: electric, magnetic and electromagnetic methods applied to cultural heritage. J Appl Geophys 41(2–3):1–180

Peacock DPS (1968) Petrological study of certain Iron Age pottery from western England. Proc Prehist Soc 34:414–427

Peacock DPS (1970) The scientific analysis of ancient ceramics: a review. World Archaeol 1:375–389

Pearsall DM (1989) Phytolith analysis. Paleoethnobotany: a handbook of procedures. Academic Press, London, pp 311–438

Pearsall DM, Trimball M (1984) Identifying past agricultural activity through soil phytolith analysis: a case study from the Hawaiian Islands. J Archaeol Sci 11:119–133

Pearson CE, Pearson SG (1989) Cultural resources reconnaissance of construction project areas of Wassaw National Wildlife Refuge, Georgia, vol 1. University of Georgia, Athens

Pearson CE, Kelley DB, Weinstein RA, Gagliano SA (1984) Archaeological investigations of the outer continental shelf: a study within the Sabine River Valley, Offshore Louisiana and Texas. OCS Study, MMS 86–0119. US Department of the Interior. Minerals Management Service, New Orleans

Penck A, Brückner E (1909) Die Alpen im Eiszeitalter, vol 2. Tauchnitz, Leipzig

Perkins D, Henke KR (2000) Minerals in thin section. Prentice Hall, Upper Saddle River, NJ

Perryman M (1964) Georgia petroglyphs. Archaeology 17(1):54–56

Petrie WMF (1899) Sequences in prehistoric remains. J Anthropol Inst 29:295–301

Petrie WMF (1904) Methods and aims in archaeology. Macmillan, London

Petrie WMF (1920) Prehistoric Egypt. British School of Archaeology in Egypt. London

Pettijohn FJ (1975) Sedimentary rocks. Harper and Brothers, New York

Pettijohn FJ, Potter PE, Siever R (1973) Sand and sandstone. Springer, Berlin Heidelberg New York

Philpotts AR (1989) Petrography of igneous and metamorphic rocks. Prentice Hall, Englewood Cliffs

Picon M, Carre C, Cordoliani ML, Vicky M, Hernandez JC, Mignard MG (1975) Composition of the La Granfensengue, Banassac and Moutans Terra Sigillata. Archaeometry 17:191–199

Picouet P, Magnetti M, Pipponnier P, Schroerer M (1999) Cathodluminescence spectroscopy of quartz grains as a tool for ceramic provenance. J Archaeol Sci 23(4):619–632

Pike SH (1995) An investigation of marble heterogeneity as a tool for characterizing ancient marble quarries on Mount Pentelikon, Attica, Greece. Geological Society of America, Abstracts Geological Society of America 27, no 6, pp 137

Piperno D (1988) Phytolith analysis: an archaeological and geological perspective. Academic Press Inc, San Diego, pp 47–49

Pollard AM (ed) (1999) Geoarchaeology: exploration, environments and resources. Geological Society Special Publication No 165. Geological Society, Bath

Pollard AM, Heron C (1996) Archaeological chemistry. The Royal Chemical Society, Cambridge

Potts PJ, Webb PC, Watson JS (1985) Energy dispersive X-ray fluorescence analysis of silicate rocks; comparison with wavelength-dispersive performance. Analyst 110:507–513

Powers AH (1989) Great expectations: a short historical review of European phytolith systematics. In: Mulholland SC, Rapp G Jr (eds) Phytolith systematics: emerging issues. Plenum Press, New York, pp 15–35

Preston D (1997) The lost man. The New Yorker, June 16, pp 70–81

Pretola JP (2001) A feasibility study using silica polymorph ratios for sourcing chert and chalcedony lithic materials. J Archaeol Sci 28:721–739

Price TD (ed) (1989) The chemistry of prehistoric human bone. Cambridge University Press, Cambridge

Ramseyer D (1992) Cites lacustres. Edition Du Cedarc, Treignes

Ramseyer K, Fischer J, Matter T, Eberhardt P, Geiss J (1989) A cathodluminescence microscope for low intensity luminescence. J Sediment Petrol 59:619–622

Rapp G Jr, Gifford JA (1985) Archaeological geology. Yale University Press, New Haven

Rapp G Jr, Hill CL (1998) Geoarchaeology. Yale University Press, New Haven

Rapp G Jr, Albns J, Henrickson E (1984) Trace element discrimination of discrete sources of copper. In: Lambert JB (ed) Archaeological chemistry III. American Chemical Society, Washington, DC, pp 273–293

Reading HG, Levell BK (1996) Controls on the sedimentary rock record. In: Reading HG (ed) Sedimentary environments: processes, facies and stratigraphy, 3rd edn. Blackwell Science, London

Reagan MJ, Rowlett RM, Garrison EG, Dort W Jr, Bryant V Jr, Johannsen CJ (1978) Flake tools stratified below Paleo-Indian artifacts. Science 200:1272–1275

Reed SJB (1996) Electron microprobe analysis and scanning electron microscopy in geology. Cambridge University Press, Cambridge

Renfrew C, Bahn P (1996) Archaeology: theories, methods and practice, 2nd edn. Thomas and Hudson, New York

Rietveld HM (1969) A profile refinement method for nuclear and magnetic structures. J Appl Crystallogr 2:65-71

Rich FJ (1999) A report on the palynological characteristics of the brown coal samples from the Ennis Mine. southeastern section. Geological Society of America Field Guide, pp 24-25

Rice PM (1987) Pottery Analysis: a sourcebook. University of Chicago Press, Chicago

Rigler M, Longo W (1994) High voltage scanning electron microscopy theory and applications. Microsc Today 94(5):2

Rinaldo A, Dietrich WE, Rigon R, Vogel GK, Rodriguez-Iturbe I (1995) Geomorphological signatures of varying climate. Nature 374:632-635

Ritter DF, Kochel RC, Miller JR (1995) Process geomorphology, 3rd edn. Wm C Brown, Dubuque

Robbiola L, Fiaud C (1992) Apport de l'analyse statistique des produits de corrosion a la compréhension des processus de dégradation des bronzes archéologiques. Rev Archeometrie 16:109-119

Robinson JW (1995) Undergraduate instrumental analysis, 5th edn. Marcel Dekker, New York

Rockwell TK (2000) Use of soil geomorphology in fault studies. In: Quaternary geochronology: methods and applications. American Geophysical Union, Washington, DC

Roper DC (1990) Protohistoric Pawnee hunting in the Nebraska sand hills: archeological investigations at two sites in the Calamus Reservoir: a report to US Department of the Interior, Bureau of Reclamation, Great Plains Region. US Dept of Interior, Billings, MT

Rovner I (1983) Major advances in archaeobotany: archaeological uses of phytolith analysis. In: Schiffer MB (ed) Advances in archaeological method and theory, 6. Academic Press, New York

Rovner I (1988) Macro-and micro-ecological reconstruction using plant opal phytolith data from archaeological sediments. Geoarchaeology 3:155-163

Russ JC, Rovner I (1989) Stereological identification of opal phytolith populations from wild and cultivated *Zea*. Am Antiquity 54 (4):784-792

Rychner V, Kläntschi N(1995) Arsenic, nickel et antimoine. Cahiers d'Archéologie Romande No.63, Tome 1, Lausanne

Salvidor A (1994) International stratigraphic guide: a guide to stratigraphic classification, terminology and procedure. IGUS, Trondheim, Norway

Sasaki Y (1989) Two-dimensional joint inversion of magnetotelluric and dipole-dipole resistivity data. Geophysics 54:254-262

Sauter MR (1976) Switzerland from earliest times to the Roman Conquest. Thames and Hudson, London

Sayre EV, Dodson RW (1957) Neutron activation study of Mediterranean potsherds. Am J Archaeol 61:135-141

Schiffer MB (1987) Formation processes of the archaeological record. University of New Mexico Press, Albuquerque

Schuldenrein J (1991) Coring and the identity of cultural-resource environments: a comment on Stein. Am Antiquity 56:131-137

Schuldenrein J (1995) Geochemistry, phosphate fractionation, and the detection of activity areas at prehistoric North American sites. Pedological perspectives in archaeological research. Soil Science Society of America Special Publication 44. Madison, WI

Schwartz HP, Liritzis Y, Dixon A (1980) Absolute dating of travertines from the Petralona Cave, Khalkidiki Penisula, Greece. Anthropos 7:152-173

Scollar I, Weidner B, Segeth K (1986) Display of archaeological magnetic data. Geophysics 51:623-634

Scollar IB (1990) Archaeological prospecting, image processing, and remote sensing. Cambridge University Press, New York

Scott DA (2001) The application of scanning X-ray fluorescence microanalysis in the examination of cultural materials. Archaeometry 43:475-482

Sealy JC, Van der Merwe NJ (1985) Isotopic assessment of human diets in the southwestern Cape, South Africa. Nature 315:138-140

Seidov D, Maslin MA (1999) Collapse of the North Atlantic deep water circulation during the Heinrich Events. Geology 27:23–26

Shackley ML (1975) Archaeological sediments: a survey of analytical methods. Wiley, New York

Shackley MS (1998) Gamma rays, X-rays and stone tools: some recent advances in archaeological geochemistry. J Archaeol Sci 25:259–270

Shane LCK (1992) Palynological procedures (draft). University of Minnesota, Minneapolis

Sharma PV (1997) Environmental and engineering geophysics. Cambridge University Press, Cambridge

Shepard AO (1936) The technology of Pecos pottery. In: Kidder AV, Shepard AO (eds) Glaze-paint, culinary and other wares. The pottery of Pecos, vol II. Yale University Press, New Haven, pp 389–588

Shepard AO (1942) Rio Grande glaze-paint ware. A study illustrating the place of ceramic technological analysis in archaeological research. Publication 528. Carnegie Institution of Washington, Washington, DC

Shepard AO (1954)Ceramics for the archaeologist. Carnegie Institution of Washington Publication 609. Carnegie Institution, Washington, DC

Shepard AO (1966) Rio Grande glaze–paint pottery: a test for petrographic analysis. In: Matson FR (ed) Ceramics and man. Viking fund publications in anthropology, no 41. Wenner Gren Foundation, New York, pp 62–87

Shopov YY, Ford DC, Schwarz HP (1994) Luminescent microbanding in speleothems: high resolution chronology and paleoclimate. Geology 22:407–410

Shuter E, Teasdale WE (1989) Techniques of water-resources investigations of the United States Geological Survey, Chapter F1: application of drilling, coring, and sampling to test holes and wells. US Geological Survey, Washington, DC

Sieveking de G, Bush P, Ferguson J, Craddock PT, Hughes MJ, Cowell MR (1972) Prehistoric flint mines and their identification as sources as raw materials. Archaeometry 14:151–176

Sigurdsson H (1999) Melting the earth: the history of ideas on volcanic eruptions. Oxford University Press, New York

Sinclair PJJ (1991) Archaeology in eastern Africa: An overview of current chronological issues. J Afr Prehist 32:179–219

Singer MJ, Janitsky P (eds) (1986) Field and laboratory procedures used in a soil chronosequence study. US Geological Survey Bulletin 1648. US Geological Survey, Washington, DC

Singleton KL, Odell GH, Harris TM (1994) Atomic absorption spectrometry analysis of ceramic artefacts from a protohistoric site in Oklahoma. J Archaeol Sci 21:343–358

Sjöberg A (1976) Phosphate analysis of anthropic soils. J Field Archaeol 3:447–454

Sneath PHA, Sokal RR (1973) Numerical taxonomy: the principles and practice of numerical classification. WF Freeman, San Francisco

Soil and Plant Analysis Council Inc (2000) Soil analysis: handbook of reference methods. CRC Press, Boca Raton

Soil Survey Division Staff (1993) Soil survey manual. USDA Handbook 18. US Dept of Agriculture, Washington, DC, p 138

Sokal RR, Rohlf FJ (eds) (1981) Biometry, 2nd edn. WH Freeman, New York

Spindler K (1994) The man in the ice. Harmony, New York

Starr JL, Parkin TB, Meisinger JJ (1995) Influence of sample-size on chemical and physical soil measurements. Soil Sci Soc Am J 59(3):713–719

Stein JK (1986) Coring archaeological sites. Am Antiquity 51:505–527

Stein JK (1988) Interpreting sediments in cultural settings. In: Stein JK, Farrand WR (eds) Archaeological sediments in context. Center for the Study of Man, Orono, ME, pp 5–19

Stein JK (1991) Coring in CRM and archaeology: a reminder. Am Antiquity 56:131–137

Stein JK (1993) Scale in archaeology, geosciences, and geoarchaeology. In: Stein JK, Linse AR (eds) Effects of scale on archaeological and geoscientific perspectives. Geological Society of America Special Paper 283. Geological Society of America, Boulder, CO

Stein JK, Farrand WR (eds) (1998) Archaeological sediments in context. Center for the Study of Early Man, Orono, ME

Stein JK, Linse AR (eds) (1993) Effects of scale on archaeological and geoscientific perspectives. Geological Society of America Special Paper 283. Geological Society of America, Boulder, CO

Stewart JD, Fralick P, Hancock RGV, Kelley JH, Garrett EM (1990) Petrographic analysis and INAA geochemistry of prehistoric ceramics from Robinson Pueblo, New Mexico. J Archaeol Sci 17:601–625

Stiner MC, Goldberg P (2000) Long gone or never was? Infrared and macroscopic data on bone diagenesis in Hayonim Cave (Israel). J Human Evol 38(3):A29–A30

Stoltman JB (1989) A quantitative approach to the petrographic analysis of ceramic thin sections. Am Antiquity 54:147–160

Stoltman JB (1991) Ceramic petrography as a technique for documenting cultural interaction: an example from the upper Mississippi Valley. Am Antiquity 56(1):103–120

Stone LM (1974) Fort Michlimackinac 1715–1781: an archaeological perspective on the revolutionary frontier. Publications of the Museum. Michigan State University, East Lansing

Stos-Gale ZA (1995) Isotope archaeology-a review. In: Beavis J, Barker K (eds) Science and site. Occasional Paper 1. Bournemouth University, Bournemouth

Streckeisen A (1979) Classification and nomenclature of volcanic rocks, lampropyres, carbonatites, melititic rocks: recommendations and suggestions of the IUGS Subcommision or the systematics of igneous rocks. Geology 7:331–335

Struve GA (1835) De silica in plantis nonnullis. PhD Diss, University of Berlin, Berlin

Szymanski JE, Tsourlos P (1993) The resistive tornography technique for archaeology: and introduction and review. Archaeol Polonia 31:5–32

Tarbuck EJ, Lutgens FT (1987) The earth, 2nd edn. Merrill Publishing, Columbus, OH

Thomas DH (1976) Figuring anthropology. Holt, Rinehart and Winston, New York

Thomas DH (1987) The archaeology of Mission Santa Catalina de Guale. Part I. Search and Discovery. vol 63, part 2. Anthropological Papers of the American Museum of Natural History, New York

Thomas DH (1998a) St. Catherines: an island in time. Georgia Endowment for the Humanities, Atlanta

Thomas DH (1998b) Archaeology, 3rd edn. Harcourt Brace, Fort Worth

Tite M, Mullins C (1969) Electromagnetic surveying: a preliminary investigation. Prospezioni Archeol 4:95–124

Tite MS (1972) Methods of physical examination in archaeology. Seminar Press, London

Tite MS (1992) The impact of electron microscopy on ceramic studies. In: Pollard AM (ed) New developments in archaeological science. Oxford University Press, London, pp 111–131

Tite MS, Linington RE (1975) Effects of climate on the magnetic susceptibility of soils. Nature 256:565–566

Toom DL, Kvamme KL (2002) The "Big House" at Whistling Elk village (39Hu42): geophysical findings and archaeological truths. Plains Anthropol 47(180):5–16

Topographic Science Working Group (1988) Topographic science working group report to the land processes branch. Earth Science and Applications Division, NASA Headquarters, Washington, DC

Tripp AC, Hohmann GW, Swift CM Jr (1984) Two-dimensional resistivity inversion. Geophysics 49:1708–17

Tukey JW (1977) Exploratory data analysis. Addison-Wesley, Reading, MA

Turner G, Siggins AF, Hunt LD (1993) Groung penetrating radar – will it clear the haze at your site? Explor Geophys 24:819–832

Twiss PC (2001) A curmudgeon's view of grass phytolithology. In: Meunier JD, Colin F (eds) Phytoliths: applications in earth sciences and human history. AA Balkema, Lisse

Twiss PC, Suess E, Smith RM (1969) Morphological classification of grass phytoliths. Proc Soil Sci Soc Am 33:109–115

Tykot RH (1996) Mediterranean islands and multiple flows: the sources and exploitation of Sardinian obsidian. In: Shackley MS (ed) Methods and theory in archaeological obsidian studies. Plenum, New York

Tzavella-Evjen H (1985) Lithares: an early Bronze Age settlement on Boeotia. Occasional Paper 15. Institute of Archaeology. University of California, Los Angeles, CA

Tzavella-Evjen H, Syropoulos T (1973) Lithares: an early Bronze Age settlement near Thebes. Archaiol Analekta Antheon 6:371–375

Ure AM (1991) Atomic absorption and flame emission spectrometry. In: Smith KA (ed) Soil analysis: modern instrumental techniques, 2nd edn. Marcel Dekker Inc, New York, pp 1–62

Vanderford CF (1897) The soils of Tennessee. Bull Tenn Agric Extension Stn 10:31–139

Van der Merwe NJ (1992) Light stable isotopes and the reconstruction of prehistoric diets. New developments in archaeological science. Oxford University Press, Oxford, pp 247–264

Van der Plas L, Tobi AC (1965) A chart for judging the reliability of point counting results. Am J Sci 263:87–90

Vaughn SJ (1990) Petrographic analysis of the early Cycladic wares from Akrotiri, Thera. In: Hardy DA (ed) Thera and the Aegean World III, vol 1. Archaeology. Thera Foundation, London, pp 470–487

Vaughn SJ (1991) Material and technical characterization of Base Ring Ware: a new fabric technology. In: Barlow JA, Bolger DL, Kling B (eds) Cypriot ceramics: reading the prehistoric record. University Museum Monograph 74. University of Pennsylvania, Philadelphia, pp 199–130

Velde B, Druc IC (1998) Archaeological ceramic materials. Springer, Berlin Heidelberg New York

Verhoogen J (1969) Magnetic properties of rocks. Geophys Monogr 13:627–633

Vineyard JD, Feder GL (1988) Springs of Missouri. Missouri Geological Survey and Water Resources, Rolla

Vitali W, Franklin UM (1986) New approaches to the characterization and classification of ceramics on the basis of their elemental composition. J Archaeol Sci 13:161–170

Vitousek PM, Chadwick OA, Crews TE, Fownes JA, Hendricks DM, Herbert D (1997) Soil and ecosystem development across the Hawaiian Islands. GSA Today 7:1–8

Wagner GA (1998) Age determination of young rocks and artifacts. Springer, Berlin Heidelberg New York

Walderhaug O, Rykkje J (2000) Some examples of the affect of crystallographic orientation on cathodluminescence colors of quartz. J Sediment Res 70:545–548

Waltham D (1994) Mathematics: a simple tool for geologists. Chapman and Hall, London

Ward G (1974) A systematic approach to the definition of sources of raw material. Archaeometry 16:55–63

Warren P (1988) Crete: the Minoans and their gods. In: Cunliffe B (ed) Origins. The Dorsey Press, Chicago

Warrick AW, Meyers DE, Nielsen DR (1986) Geostatistical methods applied to soil science. Methods of soil analysis. Part I. Agronomy monograph No 9, 2nd edn. Soil Society of America, Madison, WI

Wascher HL, Alexander JD, Ray BW, Beavers AH, Odell RT (1960) Characteristics of soils associated with glacial tills in northeastern Illinois. Agricultural Station, Bulletin No 665. University of Illinois, Champaign-Urbana

Waters MR (1992) Principles of geoarchaeology: a North American perspective. University of Arizona Press, Tucson

Weiner S, Goldberg P, Bar-Yosef O (1993) Bone preservation in Kebara Cave, Israel using on-site Fourier transform infrared spectroscopy. J Archaeol Sci 20(6):613–627

Weiss NA, Hassett MJ (1993) Introductory statistics, 3rd edn. Addison-Wesley, New York

Wendorf F, Close AE, Schild R (1987) A survey of the Egyptian radar channels: example of applied archaeology. J Field Archaeol 14:43–63

Wendorf F, Krieger AD, Albritton CC, Stewart TD (1955) The midland discovery. University of Texas Press, Austin

Weymouth JW (1986) Geophysical methods of archaeological site surveying. In: Schiffer MB (ed) Advances in archaeological method and theory, vol 9. Academic Press, Orlando

Weymouth JW (1996) Digs without digging, exploring archaeological sites with geophysical techniques. Geotimes 41:16–19

Weymouth JW, Higgins R (1985) Geophysical surveying of archaeological sites. In: Rapp G, Gifford J (eds) Archaeological geology. Yale University Press, New Haven

Whitbread IK (1989) A proposal for the systematic description of thin sections towards the study of ancient ceramic technology. In: Maniatis Y (ed) Archaeometry Proceedings of the 25th International Symposium. Elsevier, Amsterdam, pp 127–138

Whittig LD, Allardice WR (1986) X-Ray diffraction techniques. In: Klute A (ed) Methods of soil analysis, part I, 2nd edn. American Society of Agronomy, Madison, WI, pp 331–362

Williams DF (1983) Petrology of ceramics. In: Kempe DRC, Harvey AP (eds) The petrology of artefacts. Clarendon Press, Oxford, pp 301–329

Williams HF, Turner J, Gilbert CM (1955) Petrography. WH Freeman, San Francisco

Williams JM (1984) A new resistivity device. J Field Archaeol 11:110–114

Williams JM, Shapiro G (1982) A search for the eighteenth century village at Michlimackinac: a soil resistivity survey. Archaeological completion report series, no 4. Mackinac Island State Park Commission, Michigan

Williams MAJ (1984) Late quaternary prehistoric environments in the Sahara. In: Clark JD, Brandt SA (eds) From hunters to farmers. University of California Press, Berkeley, CA, pp 74–83

Williams-Thorpe O, Potts PJ, Webb PC (1999) Field-portable non-destructive analysis of lithic archaeological samples by X-ray fluorescence instrumentation using a mercury iodide detector: comparison with wave-length dispersive XRF and a case study in British stone axe provenancing. J Archaeol Sci 26(2):215–237

Wilson JT (1966) Did the Atlantic close and then re-open? Nature 211:676–681

Winiger J (1981) Spielzug und Seeufersiedlungen. Helvetia Archeol 45/48:209–217

Winkler EM (1994) Stone in architecture. Springer, Berlin Heidelberg New York

Wiseman J (1996) Wonders of radar imagery: glimpses of the ancient world from space. Archaeology 49:14–18

Wood WR, McMillan RB (eds) (1975) Prehistoric man and his environments: a case study in the Ozark Highlands. Academic Press, New York

Wynn JC (1987) Penrose conference report on archaeology and geology, 7–11 December 1986, St. Simons Island. US Geological Survey Administrative Report, US Geological Survey, Reston, VA, p 35

Wynn JC (1988) Titanium geophysics-the application of induced polarization to sea-floor mineral exploration. Geophysics 53:386–401

Wynn JC (1990) Applications of high-resolution geophysical methods to archaeology. In: Lasca NP, Donahue J (eds) Archaeological geology of North America. Centennial special volume 4. Geological Society of America, Boulder, CO

Yaalon DH (1976) Calgon no longer suitable. Soil Sci Soc Am J 40:333

Yacobi BG (1994) Cathodoluminescence in microanalysis of solids. In: Yacobi BG, Holt DB, Kazmerski LL (eds) Microanalysis of solids. Plenum Press, New York

Yardley BWD, Mackensie WS, Guilford C (1990) Atlas of metamorphic rocks and their textures. Longman, Harlow, UK

Index

Printed in the United States
131985LV00003BA/7-12/A

9 783540 438229